BECOMING BIOSUBJECTS
Bodies. Systems. Technologies.

Becoming Biosubjects examines the ways in which the Canadian govern-ment, media, courts, and everyday Canadians are making sense of the challenges being posed by biotechnologies. The authors argue that the human body is now being understood as something that is fluid and without fixed meaning. This has significant implications on both how we understand ourselves and how we see our relationships with other forms of life.

Focusing on four major issues, the authors examine the ways in which genetic technologies are shaping criminal justice practices, how policies on reproductive technologies have shifted in response to biotechnolo-gies, the debates surrounding the patenting of higher life forms, and the Canadian (and global) response to bioterrorism. Regulatory strategies in government and the courts are continually evolving and are affected by changing public perceptions of scientific knowledge. The legal and cultural shifts outlined in *Becoming Biosubjects* call into question what it means to be a Canadian, a citizen, and a human being.

NEIL GERLACH is an associate professor in the Department of Sociology and Anthropology at Carleton University.

SHERYL N. HAMILTON is an associate professor in the Department of Law and the School of Journalism and Communication at Carleton University.

REBECCA SULLIVAN is an associate professor in the Faculty of Communication and Culture at the University of Calgary.

PRISCILLA L. WALTON is a professor in the Department of English at Carleton University.

NEIL GERLACH, SHERYL N. HAMILTON,
REBECCA SULLIVAN, AND PRISCILLA
L. WALTON

BECOMING BIOSUBJECTS
Bodies. Systems. Technologies.

UNIVERSITY OF TORONTO PRESS
Toronto Buffalo London

ISBN 978-0-8020-9983-9 (cloth)
ISBN 978-0-8020-9683-8 (paper)

Printed on acid-free and 100% post-consumer recycled paper with vegetable-based inks

---

Library and Archives Canada Cataloguing in Publication

Becoming biosubjects: bodies, systems, technologies / Neil Gerlach ... [et al.].

(Cultural studies series)
Includes bibliographical references and index.
ISBN 978-0-8020-9983-9 (bound)   ISBN 978-0-8020-9683-8 (pbk.)

1. Biotechnology – Social aspects – Canada. 2. Genetic engineering – Social
aspects – Canada. 3. Human body. 4. Forensic genetics – Technique.
5. Criminal justice, Administration of – Canada. 6. Reproductive technology –
Government policy – Canada. 7. Biotechnology – Canada – Patents.
8. Bioterrorism – Canada – Prevention. 9. Bioterrorism – Prevention.
I. Gerlach, Neil, 1963– II. Series: Cultural studies series (Toronto, Ont.)

TP248.23.B43 2011   303.48'3   C2010-900559-7

University of Toronto Press acknowledges the financial assistance to its publishing program of the Canada Council for the Arts and the Ontario Arts Council.

University of Toronto Press acknowledges the financial support for its publishing activities of the Government of Canada through the Book Publishing Industry Development Program (BPIDP).

This book has been published with the help of a grant from the Canadian
Federation for the Humanities and Social Sciences, through the Aid to
Scholarly Publications Program, using funds provided by the Social Sciences
and Humanities Research Council of Canada.

Contents

Acknowledgments

We would like to extend our gratitude to the Social Sciences and Humanities Research Council of Canada and to the Canadian Federation of the Humanities and Social Sciences' Aid to Scholarly Publication Program for their generous support of the research and publication of this book. The staff at the University of Toronto Press were enthusiastically behind this project from the beginning. Special thanks to Siobhan McMenemy, Ryan Van Huijstee, Frances Mundy, Patricia Simoes, Leah Connor, and Ian MacKenzie. Our reviewers provided insightful comments and worthy criticism that helped improve this book. We are grateful to them for taking such care in their assessments.

Neil Gerlach thanks Emily Truman for her invaluable research assistance on this project. I would like to thank my co-authors for their good humour, enthusiasm, and stimulating discussions. As well, I thank my colleagues in the Department of Sociology and Anthropology at Carleton University for their ongoing support. Finally, I would like to thank my wife, Sheryl, and our daughter, Brigitte, for their constant inspiration.

Sheryl Hamilton would like to thank the Canada Research Chairs program, Carleton University, and in particular, the Office of the Vice-President Research and International, the dean of the Faculty of Public Affairs, and my friends and colleagues in the Department of Law and the School of Journalism and Communication. Paula Romanow and Emily Truman provided unflagging, intelligent, and invaluable research assistance. I would like to thank my co-authors for their energy, patience, good humour, and intellectual stimulation over the years. Last, but never least, I would like to thank my husband, Neil, and our daughter, Brigitte, who ground me and without whose love and support none of this would be any fun.

Rebecca Sullivan thanks the University of Calgary's Research Excellence Envelope for funding this project in its nascent stage. I am indebted to Edna Einsedel, who provided mentoring at a crucial time in my career. Gwendolyn Blue was generous with her time and her ideas, providing much-needed collegial and intellectual support. Jaime Wood, Nils Olsen, Darren Blakeborough, and Carol Neuman were all able research assistants along the way, and I thank them for their efforts on my behalf. My co-authors deserve my deepest gratitude for seeing me through major personal life changes as this project progressed. Finally, I want to give a special thank you to Bart Beaty and our son, Sebastian, who make everything possible for me.

Percy Walton would like to thank Chris Eaket for his inestimable research skills. I am indebted to my co-authors, who made this project a wonderful and exciting experience. Finally, my deepest thanks to Michael, to whom I owe it all.

BECOMING BIOSUBJECTS
Bodies. Systems. Technologies.

CHAPTER ONE

Introduction

The new genetics will cease to be a biological metaphor for modern society and will become instead a circulation network of identity terms and restriction loci, around which and through which a truly new type of autoproduction will emerge.

Rabinow (1996, 99)

Andrew Niccol's stylish 1997 feature, *Gattaca*, provides a look into a future where, as its tag line suggests, 'there is no gene for the human spirit.' The film depicts the life of a 'degenerate,' or a non-genetically enhanced man (played by Ethan Hawke), who attempts to prove that he can compete with those genetically 'superior' to himself. Barred from even attempting to do so because of his 'degenerate' status, Hawke's character is forced to 'pass,' and he does so by locating a genetically enhanced but now paralyzed partner (played by Jude Law). The two reach an agreement – Hawkes's character will care for Law's in exchange for Law's bodily fluids, which he needs to gain entry into the world from which he is otherwise excluded. In its cold genetic determinism, *Gattaca* portrays a society where one's life path is determined by a genetic screening conducted at birth, where parents team with private science to construct better babies, and where the state and capital strive to minimize the risks produced by the non-genetically enhanced.

From a different perspective, Roger Spottiswoode's *Sixth Day* (2000) presents Arnold Schwarzenegger as a man confronted with his own clone, and his distaste for what he perceives as this biological 'perversion' of himself. Ultimately, he comes to a tenuous acceptance of his 'other,' and the two team to fight a conspiracy generated by corporate greed and the desire to live forever. Clearly, both *Gattaca* and *The Sixth Day* figure biotechnology as a source of human devastation and social degradation, and sit in stark contrast to the claims of scientists and governments, who are often featured on news broadcasts or in press interviews, actively resisting such popular representations. These experts attempt to counter the science fiction imagery with promises of a host of new, experimental therapies designed to improve, prolong, or even create life.

Indeed, as these images suggest, biotechnology is represented typically as either the hope for humanity's salvation or the most likely source of its complete destruction. These extremes mark the ends of a continuum between fear and promise structuring the public imagination around biotechnology. Fear is solicited by framing genetically modified food-stuffs as Frankenfoods, in the predicted slide to genetically based eugenics and in the spectre of bioterrorism. Promise manifests itself in gene therapies intended to cure ailments such as Huntington's disease, in the quest for replacement organs, and in genetically based techniques for assisted reproduction.

What is striking about these representations – whether found in the press releases of scientific organizations, industry advertising, the participants in government hearings, films, fiction, or the appeals of transnational NGOs – is that they are all ultimately forms of fiction. There is no genetic cure for Huntington's disease; the impacts of genetically modified foods are unknown; mature human clones have not yet been produced; and neither babies nor hearts are grown in vats. In other words, the various claims through which members of the public have access to biotechnological knowledge are not yet scientific or social facts. At best, they could be called social science fictions, or discursive frames that are used to describe the social or institutional effects of an imagined but not quite yet realized technology (Bogard 1996, 8). Social science fictions are the frames and narratives within which we locate unfamiliar, underdeveloped, or as yet unknown genetic technologies. The future possibilities of these technologies are folded seamlessly into their present description. In this way, the technology is mystified and ultimately reified, making it less amenable to critical analysis. The choice is not, then, whether we should have or use this technology, but rather, how to deal with its effects, as the social science fictional framing has rendered it already present. It is important, therefore, to analyse not only the technologies themselves, but the social science fictions in and through which we understand and talk about them.

We are using social science fiction in this way because of its refusal to privilege either of the utopian or dystopian constructions of the future technologies. Instead, it places the two together in a dynamic relationship in which new vistas of understanding and meaning-making are forged. In *Becoming Biosubjects* we suggest that a set of social science fictions are shaping the Canadian encounter with biotechnology, through narratives and imagery that conflate future possibilities and existing social realities into an unstable and yet shared, present experience.[1]

While it can be argued that such techniques as animal and plant husbandry, which have been a part of human culture for thousands of years, are also at least early precursors of biotechnology, the interests that motivate *Becoming Biosubjects* stem predominantly from those biotechnologies that intervene on living matter at the molecular level. Biotechnology thus forces an encounter with new ideas and possibilities with the potential to radically transform notions of bodies, humanness, life, and society. As a result of this incredible potential, biotechnologies both force issues of governance and make problematic resulting attempts to govern them. What is and is not a proper domain into which the state can or should intervene is not self-evident, as the categories previously used to answer those questions are in flux. Yet states and citizens alike are challenged to negotiate new boundaries of propriety, new forms of identity, and new manifestations of power. This governmental activity takes place at the nexus of the institutional, the mass mediated, and the everyday; it creates new languages and representational schema through which individuals come to a sense of shared meaning about the society in which they live and about themselves. In the following pages, we take up a number of contemporary examples that help us understand these processes in Canada, which make three broader contributions, we suggest, to the existing literature.

First, the Canadian instance is important because the bulk of literature exploring how biotechnology is both interpreted and governed concentrates overwhelmingly on the American, European, and developing world experiences (e.g., Bunton and Peterson 2005; Nelkin and Lindee 1995; Roof 2007; Shiva 1997; Turney 1998). Canada is rarely, if ever, considered, and the specificity of how biotechnological issues play out in the Canadian context goes unrecognized.[2] Interestingly, in a number of the domains considered here, the meanings, governmental strategies, and social outcomes in Canada diverge radically from those in the United States and Europe. This difference suggests a value to the study of the Canadian experience that accrues not only to the Canadian reader. Further, the social science fictions that structure the Canadian experience are still contested, and, hence, studying Canadian public cultures makes visible the relations of biopower elsewhere obscured or already resolved.[3]

Second, in literature focusing on governance and biotechnology, the focus is almost exclusively technical and normative – how to better govern – rather than how one might interpret governmental practices as ways of making meaning (Bunton and Peterson 2005; O'Mahony 1999; Thacker 2005). As a result, cultural or mediated texts are ignored as sites for the negotiation of governance. One notable example to the contrary, is José van Dijck's *ImagEnation: Popular Images of Genetics* (1998). Studying scientific, governmental, and public discourses, she offers a very useful historical typology of dominant images of the gene and how they shape both political action and politics. The first vision, she suggests, is that of the gene as alphabet or language, and is dominant through the 1950s and into the 1960s. The second is the gene as laboratory monster that may escape and pollute nature, emerging in the 1960s and continuing until the 1970s. Third, is the gene as 'master controller' or manager, responsible for regulating all other aspects of the human body, a discourse that emerges out of the explosion of sociobiology in the 1980s. Fourth and finally is the gene as a map, drawing on the imagery of the Human Genome Project. Van Dijck argues that this last image is very important because the map is an information map; that is, given the cultural status of information and science in contemporary society, genetic information moves from representation to something with ontological status.[4]

Third, the existing scholarship, while effectively mapping out the conditions of possibility for the emergence of Paul Rabinow's 'truly new type of autoproduction' does not explore that outcome in any detail. *Becoming Biosubjects* refocuses the gaze upon governmental practices and their implications for changing subjectivities. The following pages argue, in agreement with Rabinow's assertion, that the gene is now more than a metaphor, that it has the power to locate a person's position within different social fields, to alter one's perceptions of oneself, and to shape the pathways of their regulation and mobility. Consequently, a powerful new mode of subjectivity emerges in this altered field of social relations. This biosubjectivity troubles traditional modernist dualisms between natural and artificial, human and animal, private and public, and present and future. The subject is both alienated from and dependent upon a fragmented body. It is a subject outside of humanist ethics and firmly within capitalist relations. Yet it is not a determined subject; it is responsible for its own production. It is a manifestation of the encounter between present reality and future possibility. It is a subject that is always already in conversation with other late modern subjects – the entrepreneurial subject, the prudent subject, the subject under surveillance.

'Messy' Bodies: Shifting Body Ontologies

A precondition of this reformed social subject is the radical transformation of the body through bioscience and biotechnological alteration. The biotechnological body is 'messy' in two respects. The first relates to what philosopher Nicholas Agar (2004), in the context of human cloning and genetic engineering, terms the 'yuck' argument; he contends that these practices 'are wrong because they violate some deep, inchoate sense of what is right for us' (55–6). He frames his assertion more colloquially when he asks, 'Are enhancement technologies wrong because they are "yucky"?' (56). One has to think only of the injection of a deadly toxin under the skin of an aging human face in order to reduce wrinkles (complete with paralysis, subsequent sagging, and potential pain), the development of cows designed to produce human milk, or the artificial conception of a human baby to provide replacement organs for an ill sibling, in order to understand how messy the body has become at a material level. Transgenics, genetic engineering, and cloning all fundamentally alter the material composition of the human or animal body, contaminating it through 'unnatural' manipulation or cross-species interaction. Dolly, the cloned sheep, highlights the fact that biotechnological science has not yet been able to control the body's messiness effectively. All too often, the 277 prior unsuccessful attempts to clone a Dolly are forgotten, as is the fact that she died a premature death by euthanasia in early 2003 as a result of accelerated aging of her cells. While this is not an argument for the power of nature versus the power of the scientist, the resilient messiness of the human body, in particular, is thrown into relief by an increasing number of biotechnological practices.

Yet, perhaps even more importantly, the body, as a concept, has become messy. One effect of the biological intervention in the material constitution of the human body is to fundamentally destabilize our understanding of the body as a singular unit. Irma van der Ploeg (2002) suggests that the body has an intimate relationship with developing forms of technology, which results in different identity formations and embodiment experiences; in other words, there are now various 'body ontologies.' Van der Ploeg argues that the modernist body ontology has been disrupted in favour of an understanding of the body as a source of information and communication flows.[5] The anatomical/physiological body of modernity has been slowly usurped by a fluid communicative or informational body comprising bits that can be separated out and put to different use.

This study takes the notion of shifting body ontologies one step further, since the dominant body ontology of modernity can be seen as one of bodily integrity. Bodily integrity suggests that the body is a discrete unit that contains the self. Bounded by the skin, the visceral aspects of the body are confined and controlled in a kind of private space of the self, rendered invisible and thus discrete. Bodily integrity is intimately associated with a subject's capacity to exercise sovereignty over her or his body/self. Such a combination of body and self produces and sustains the human being as a political, legal, and social person. The modernist notion of bodily integrity structures intersubjective relations and is most apparent in moments of challenge, such as consent for medical care, criminal evidence collection, and reproductive issues such as forced pregnancies or abortions. Just as important, however, bodily integrity serves as a limit, a defining relationship between the citizen and the state, which historically functioned as an intimate boundary that the state could not pierce without the person's consent or an overwhelming public interest.

But bodily integrity is under assault – not only as a legal notion, but also as an ontology. The new emerging body ontology results from a number of factors, and foremost among them are biotechnological developments. Biotechnology requires a re-visioning of the body as something less than the sum of its parts. In the chapters that follow, we trace a range of sites where the human body is fragmented and rendered as divisible and permeable – all a necessary effect of the operation of biotechnologies. In this way, the person can be subject to a variety of manipulations that would have been ethically or politically jarring in the modernist body ontology. Accordingly, the 1980s debates that arose around the state's right to require a breathalyzer test for people suspected of drunk driving did not manifest in relation to the state's right to require a mouth swab for a DNA test. A further example: pregnant women are susceptible to a host of genetic screening technologies that make the previously invisible now visible and make once-private decisions about family planning a part of a larger public culture of health. As a denaturalized object or set of objects – genes, cells, proteins, organs – the body, or more accurately, elements of the body, can then be manipulated in and through a variety of social institutions and locations. The body as a whole is disrupted, resulting in a new body ontology: the 'contingent body.'

Contingent bodies are temporary and volatile; they manifest or modify, depending on particular combinations of social actors and forces. The important elements of the person become its molecularly defined components, which can then be informated, sold, killed, manipulated, reproduced, copied, or circulated along networks of exchange and knowledge production. The shift in body ontology from bodily integrity to contingent bodies de-reifies the body. It no longer stands outside of social production; the body has lost its inviolability, and, with it, its privileged status as a basis for meaning-making, political and legal rights, and ethical claims. It can be 'they,' as no necessary holism or singularity is required. Yet, at the same time, as the body loses its supreme status, bodily material such as genes and DNA are fast gaining mystical qualities. Called the 'Holy Grail,' 'the language of life,' and so on, genetic information now belongs not to the person or the person's body, but rather to a greater cosmological order altogether.[6] Hence, bodily integrity ceases to mean much more than an unnecessary, if not irritating, border to be crossed in order to arrive at new frontiers of knowledge.

This is not to suggest that contingent bodies have no relationship to meaning-making, political rights, or ethics; instead, it is to stress that that relationship is not predetermined. It emerges, situationally, wherever and whenever the particular instantiation of the body appears, including outside of biotechnological sites. Although it is produced as a result of the body's subjection to the molecular gaze, the contingent body is not simply a biotechnological object. For example, in a post-9/11 world, the immigrant's body is a contingent body. The struggle by some border detainees in the United States and Canada has highlighted the instability of, and struggle for, political and legal rights through the locus of a body that has been fragmented, de-reified, and reclassified as a result of techniques of surveillance, selection, simulation, and policing. Identity cards place the biometric indicators of risky personhood in a code that communicates certain 'truths' to officials. Sovereign subjectivity is sacrificed; rights become contingent upon what one looks like, where one comes from, who one's friends and family are – all read as signs on, or more accurately, of, the body.

This particular non-biotechnologically mediated contingent body highlights how personhood, sovereignty, and privacy are and can be manipulated by new regimes of surveillance and simulation. The remapping of body ontologies from one of integrity to one of fragmentation and contingency, then, is intricately tied to notions of public/private but also more importantly to a belief that the 'truth' of one's identity exists not on the material plane of the body, but within a microcosmic regime of genetic information. Carlos Novas and Nikolas Rose (2000) argue in favour of a form of 'somatic individuality,' in which 'bodily truths' are made knowable through genetic screening and surveillance (502). Yet this idea takes the contingent body for granted, placing it in new legal and ethical fields wherein one's DNA becomes part of an uncontested public domain of knowledge and intervention. Indeed, from strip-searching immigrants at the border to DNA dragnets, the idea that everyone is a suspect – or merely suspect – is fast becoming a dominant mode of social relations.

Biogovernance: Framing Politics

Contingent bodies force a rethinking of bodily social relations from ethical and legal perspectives, frequently enacted within governmental policy. The ontology of bodily integrity presupposed the interaction of a bounded and sovereign subject with a bounded and sovereign state. Contingent bodies pose a challenge to modernist strategies of governance in that their subjects are more difficult to identify and contain from the state's perspective. More broadly, biotechnology has proven to be a difficult subject of governance, given its complicated location in global webs of science and capitalism. At the same time, however, biotechnology provides governments with a host of new strategies and techniques to monitor its citizens and maintain social order. Those systems of surveillance and control, which Michel Foucault terms 'governmentality,' frame how biotechnologies are introduced into public culture and keep them from being dismissed as mere dystopian fantasies. It is for this reason that much of the following analysis crosses legal and governmental regimes with mediated narratives and representations. Public cultures of biotechnology are elements of biopower strategies working to create new fields of subjectivity and sovereignty. As a result, it is important to map, rather than assume, governmental interactions with contingent bodies and their associated subjects. Indeed, governmental biopower operates on two levels: the increasingly indirect relations between the state and its citizens, and the shifting obligations of subjects to govern or care for themselves.

Michel Foucault, in the first volume of *The History of Sexuality* (1986) and in several lectures given in the mid-1970s (Foucault 2003) outlines his notion of biopower to demarcate a field of rationalized governmental activity beginning at the end of the eighteenth century whereby vital characteristics of human life – health, reproduction, death, sexuality, and so on – are brought within regimes of power and governance. 'Bio-power brought life and its mechanisms into the realm of explicit calculations and made knowledge/power an agent of transformation of human life ... Modern man is an animal whose politics places his existence in question' (1986, 143). In his historical schema, it is a post-sovereign organization of power.

Rather than a mode of power that operates directly on the body, biopower operates on the notion of 'man-as-living-being' and ultimately on 'man-as-species' (Foucault 2003, 242). We can see a level of generalization and abstraction at the heart of biopower. The population emerges as both a biological and political problem (245). Life cannot be managed at the individual level, but it can be at the group level, over time. 'It is ... a matter of taking control of life and the biological processes of man-as-species and of ensuring that they are ... regularized' (246–7). In an interesting and telling distinction from sovereignty, Foucault suggests, 'Sovereignty took life and let live. And now we have the emergence of a power that we would call the power of regularization, and it, in contrast, consists in making live and letting die' (247). He clarifies the distinction: 'Power has no control over death, but it can control mortality' (248). Bodies and populations thus emerge as two poles within the operation of biopower. Life is refigured as an object of governmental power.

Various thinkers have taken up the rich but briefly articulated notion of biopower from Foucault. In an address given in late 2003, Paul Rabinow and Nikolas Rose usefully revisit the concept specifically in relation to current practices of biotechnology. They suggest that biopower designates a 'plane of actuality' that comprises a number of elements. 'The concept of biopower seeks to individuate strategies and configurations that combine a form of truth discourse about living beings; an array of authorities considered competent to speak that truth; strategies for intervention upon collective existence in the name of life and health; and modes of subjectification, in which individuals can be brought to work on themselves, under certain forms of authority, in relation to truth discourses, by means of practices of the self, in the name of individual or collective life or health' (14).

Elsewhere, Neil Gerlach (2004) has productively combined governmentality theory and notions of biopower to develop a model for understanding how governmentality is changing to address biotechnological challenges. He suggests that the resulting form of 'biogovernance' encompasses a set of management techniques aimed at tackling the risks of biotechnology in order to transform them into instruments of governance. He suggests that five interrelated processes comprise the enabling conditions of the emergent biogovernance: privatization, politicization, objectification, normalization, and responsibilization. We will revisit each component briefly here, as each of these practices appears often and in different ways in our four case studies.

Privatization prescribes the location and management rationality of genetic research. Genetic research takes place almost exclusively within the private sector, resulting in an increasingly concentrated multi-billion-dollar genomics industry that operates largely in secrecy. This secrecy is enabled by scientific and legal mechanisms such as bio-prospecting, biopatenting, and trade secrets. The consequences of such an atmosphere are threefold. First, many public institutions have adopted private sector knowledge-development models – witness the emergence of commercialization protocols in all research universities and hospitals in Canada and elsewhere. Second, the state often partners with the private sector; the most well-known example can be found in the private/public conflict over the mapping of the human genome, which was resolved in a rare instance of government and industry cooperation. Third, governments find themselves forced to rely very heavily on the advice of biotechnological experts in formulating policy. The Canadian Biotechnology Advisory Committee, constituted in 1999 to advise the Canadian government on all issues of biotechnology, consisted almost exclusively of scientific and industry representatives. As a result, the framing of biotechnological knowledge and its social impacts are always already a by-product of industry. Tellingly, the Canadian government's commercial biotechnological strategy is overseen by Industry Canada, while its regulation remains under the aegis of Health Canada.

The second precondition of biogovernance is politicization, which contains cultural dissonance and formal and informal political conflict. As a result of public reflexivity, or increased scepticism about all types of science, various interest groups seek to challenge the authority of bioscience. These conflictual processes highlight the glaring lack of an appropriate forum for negotiation and of a shared language for discussion. For example, social movements such as Greenpeace employ a language of natural rights on issues of bioprospecting, and genetically modified organisms, whereas the biotech industry and governments deploy a language of assigning responsibility for risk. The spectre of Frankenstein's monster plays out across a wide field of media and popular culture representations of biotechnology. By contrast, official scientific and governmental publications prefer cosmological images of genes and DNA to promote a sense of awe and wonder.

Unlike in the United States and Europe, Canadian politicization has been relatively muted, more likely to take place in the courts than in Parliament. Issues such as abortion, human cloning, DNA banking, and the patenting of life forms have been far less contentious in Canada than elsewhere. This renders visible two further structuring absences that mark much of the Canadian biogovernmental playing field. In different ways and in different sites, both the Canadian government and the Canadian public have been either absent from, or very slow to act in, the political debates around the ways in which biotechnologies should be implemented and governed. The Canadian government has been reluctant, or has declined outright, to legislate in key areas of biotechnology policy – genetic patenting, biotechnological reproduction issues, and the labelling of genetically modified foodstuffs. At other biogovernmental junctures, the Canadian public's response to issues that have drawn vociferous citizen protest elsewhere seems to be one of either acquiescence or apathy, apparent in the patenting of higher life forms and the implementation of a DNA databanking system for criminal offenders.

These structuring absences could be said to reflect a fundamental depoliticization in Canada, both formally and informally, in relation to certain types of biotechnology. However, they also highlight the difficulties the state faces in attempting to regulate biotechnology, a difficulty well highlighted in Canada's encounter with reproductive technologies. Arguably, this is due to the combination of moral issues seemingly at the heart of each expression of biotechnology, and the level of scientific expertise required to claim any understanding of the subject matter. Governments increasingly operate through an economic/legal frame; consequently, either moral questions must be rewritten as economic or legal, or they go unheard. While other countries, most notably the United States, have entered the moral arena of biotechnology, the Canadian government has preferred to take an incremental and even contingent approach. As Rebecca Sullivan (2005) argues elsewhere, Canadian biotechnology policy on reproduction has not rested on moral principles of life, but on the negotiated terrain of health. Such a difference has far-reaching implications for biotechnologies that directly intervene on the body. These implications are apparent in reproductive technologies, as well as in the growing field of gene patenting of manipulated life forms. Accordingly, it is worth exploring in detail how the Canadian government has parted from its international allies in establishing regulatory frames that lead to very different modes of understanding, and divergent methods of integrating biotechnology into public culture. This is not to suggest that biotechnology is ungovernable; rather, such challenges account for the uneven and inconsistent approach the Canadian state has taken to biotechnological issues.[7]

The third antecedent of biogovernance is objectification, or the production of the gene as a field of management, which includes such practices as mapping, testing, coding, banking, imaging, simulating, and representing. There are a number of implications to objectification, since new biotechnologies promise to bridge the divide between nature and culture by subjecting both to the same industrializing techniques. As discussed above, the shift in body ontology between bodily integrity and contingent bodies is an example of the broader process of objectification. In turn, from a governmental perspective, objectification also provides a means for pre-detecting human conditions and attributes that pose a genetic risk to the population at large. During the Canadian SARS crisis in Toronto in 2003, therefore, travellers from Asia, and, in particular, Asian Canadians, were marked as health risks or potential disease vectors. Their rights and the pathways of social mobility they enjoyed were altered both formally by governmental and health institution responses, and informally, through social practices of aversion and ostracism.

Fourth, biogovernance depends upon normalization. By normalization, Gerlach (2004) is identifying a cluster of practices aimed at managing public discussion of biotechnology, rendering it legitimate, normal, and secure. Rather than politicization, which occurs once a conflict has emerged in the public sphere, normalization is a strategy aimed at controlling meaning-making before it reaches the controversial stage. Hence, normalization includes specific techniques of expert and public consultation, social marketing, and legislating. Authorities attempt to produce an ethos of biotechnological optimism, the effects of which include frames for understanding social impacts, and work to limit public debate. These frames attempt to shift the public imagination from fear to promise.

The Canadian government has a long history of diffusing potential conflict in all domains of the social through consultative techniques. This has played itself out in the biogovernmental field with the Royal Commission on New Reproductive Technologies in the early 1990s, and more recently the Canadian Biotechnology Advisory Committee and its work. Through these organizations, the state creates a buffer between itself and interest groups that would otherwise engage in more direct forms of lobbying. Another unique example in Canada were the 'citizens' juries,' established to solicit public opinion on xenotransplantation (or the insertion of animal organic material into human bodies). Citizens' juries – a combination of traditional survey techniques and more in-depth, educational focus groups – resulted in an abrupt change of governmental policy from xeno research to human stem cell research. Perhaps because the results of this experimental public consultation were not to the liking of scientists, the citizens' juries have not been standardized in the policy process.[8] The benefits of commissions, advisory groups, and the like are that they satisfy demands for democratic process by almost always incorporating public consultations, which, upon closer examination, are rarely, if ever, accessed by a broad segment of the Canadian public. Yet they do serve as a governmental testing ground for different languages, imagery, and symbolism through which to express biotechnological notions and issues.

Finally, privatization, politicization, objectification, and normalization combine to produce the fifth condition of possibility for biogovernance: responsibilization. Governance becomes focused less upon the distribution of resources and concerns and more upon the work of assigning risk responsibility. In other words, responsibilization individualizes social responsibility for managing the risks of biotechnology. Increasingly, individuals are expected not to discipline themselves, but to manage themselves and the risks they might pose to the wider social good; they must do so by accessing and mobilizing the resources and expertise at their disposal in the genetic marketplace. As a case in point, fetal genetic screening has shifted from being a technological option to a virtually required process, which any parents deemed to be in a risk category for birth defects (due to age, family history, genetic predispositions, etc.) must undertake, or face social and medical approbation. The next logical step, biogovernmentally, will be for insurance regimes to be tied to the 'choice' to screen genetically (and abort appropriately, or not).

Responsibilization leads to the rethinking of what citizenship means. Of course, citizens in liberal democracies have always had rights and responsibilities to themselves and the social. The difference lies in the shift in the articulation of citizenship from a rights to a responsibility discourse. Moreover, it is not coincidental that this shift has occurred simultaneously with the rise of genetics and what Piet Strydom has called a 'civilization of the gene' (1996, 21). The nature, number, and types of responsibilities have changed as the truths of our bodies are transformed from those of integrity to fragmentation, and 'nature' becomes a problem to be solved by technology. Individuals are responsible, as Ulrich Beck (1992) has argued, to produce their own biography. In other words, identity-forming institutions that used to be instrumental in developing the self no longer serve this function to the same degree. For instance, religion, class, family, political community, neighbourhood, and so on do not offer sufficient templates or resources for self-definition. Consequently, the individual becomes a 'world shaper,' someone wholly responsible for his or her life chances, mobilizing resources as diverse as self-help literature and flu shots.

The temporality of responsibility has also changed. We used to be responsible for the actual dangers we might pose to our neighbour or community; now we are held accountable for the risk we might pose to others. This takes place, of course, in the broader context of the downloading of certain types of responsibility from the state to the individual, and has produced some interesting social effects. As Rose (2001) describes, a community has evolved as a private response to Tay-Sachs disease, given the inability (or inefficacy) of public institutions to devote sufficient resources to a cure. Interestingly, this example also highlights the new forms of solidarity that are provoked by responsibilization practices. These practices affect the kinds of experts that emerge and the objects to which their expertise is oriented. For example, there has been a deprivileging of criminal profiling, understood as an inexact science, in favour of the search for genetic guarantees of violent masculinity, all in the hunt for the 'criminal gene.'

When considering biogovernance, questions arise about the beginning of the new 'age of biological control,' pronounced by Ian Wilmut, the creator of Dolly. If it has begun, how is it proceeding? As the discussion of contingent bodies suggests, it is the dismantling of citizens' rights over their bodies that has made possible a culture of responsibilization, in which bodily components are called upon to reveal themselves in service to the state, and an increasingly normalized notion of what counts as the social good. Biotechnologies, in this sense, become both the means and the ends of their own justifications. In other words, the ability to intervene upon and fragment the body has led to new sets of regulatory and social conventions, which insist that their observance is for the good of society. The practices of biogovernance that enact these new body ontologies have been predicated upon techniques of surveillance and simulation, or, more simply, vision.

Ways of Seeing: An Epistemology of Representation

Along with the emergence of the ontologically contingent body and the shifts in the objects and techniques of governing biotechnological subjects come very specific ways of knowing and seeing, which arise, in part, from the struggle to apprehend the biotechnological subject. Novas and Rose attempt to capture this epistemological quandary in their notion of the molecular optic. They assert that people live in a molecular optic where 'life is now imagined, investigated, explained, and intervened upon at a molecular level' (2000, 487). A privileged set of apprehension techniques is emerging, all indebted to visioning. Increasingly, one recognizes one's self as a subject through practices of looking, watching, and seeing (see Dumit 2004). From surveillance to screening to modelling, the capacity to see, and then represent, has ceased to be merely representational and has become an act of creation. While scientists have long privileged vision as the unobtrusive, objective path to knowledge, critics like Evelyn Fox Keller argue that the biological gaze is far from neutral. Rather, it is the anticipatory step toward intervention and manipulation and therefore makes its object immediately technologically mediated and malleable (2000, 120). The materiality of vision, therefore, has significant consequences for the organization of biotechnological knowledge.

Catherine Waldby (2000) suggests that the problem of 'bodily opacity' has been a longstanding obsession for medical science: 'The entire anatomical enterprise, and its long historical project to bring medicine under the sign of science, is organized around the visualization of pathology in the corporeal interior' (24). While clearly the problem of seeing is longstanding within the scientific enterprise's engagement with the human body and subject, molecular visioning has taken this problem to a new level of abstraction. Van Dijck comes closer to describing the biotechnological situation in her claim that the Human Genome Project does more than simply represent DNA sequences; it translates them into an informational realm, into digital codes, all with very powerful knowledge effects: 'This digital 'registration,' however, is not merely a representation of molecular language, but more like a translation into a different dimension. Through the inscription of 'DNA-language' in digital data, the body is turned into a sequence of bits and bytes whose function is no longer exclusively representational. Besides being a representation of an organic body, the digital genome data now also function as the material inscription of a 'model' body. Digital data can be recorded and rearranged to form an 'ideal sequence,' the gold standard which has no reference in reality' (124). As van Dijck contends, the individual has an increasingly intimate relationship with the visioning technologies that enable one to know, and therefore produce, one's self.

Even so, the biological gaze is not simply a practice of technological mediation or of mere surveillance. It is also a retraining of the imagination to conceive of the body's inner spaces as the final frontier. Scholars from Nelkin and Lindee (1995) to Sawchuk (2000) have insisted that the transformation of the body from physiological object to anatomical and now molecular landscape has enormous implications for how people know their bodies and their selves. Elsewhere, Sheryl N. Hamilton (2003) has urged that greater attention be paid to the ways in which science is imagined, in order to grasp the significance of biotechnology's visionary future. Ever since the appearance of the now famous photo of Watson and Crick, gazing in wonder at a gigantic model of the DNA molecule, the inner spaces of the body have been effectively turned into gargantuan vistas of biotechnological promise, much like images of outer space in previous decades. These techniques of surveillance and simulation have produced, according to William Bogard, a fictive geography of new biotechnologically mediated subjectivities, from cyborgs to clones (1996, 6). Hence the ontology of contingent bodies cannot be separated from how one looks at one's body as awe-inspiring fragments, which can be dissembled and reassembled to uncover the truth about the self.

Contemporary truth regimes continue to privilege vision – not the vision of the unaided eye but technologically mediated vision at the molecular level. It is this molecular gaze that has the ability to apprehend the truth. This is demonstrated very effectively in a 2004 episode of the forensic crime drama *CSI: New York* entitled 'Creatures of the Night.' A young woman, raped and severely beaten, wanders out of Central Park. When she is later interviewed by police, she cannot remember the assault. Forensic technicians find considerable evidence from the crime in the form of a geranium leaf on the victim's clothing, indicating where the crime occurred, a boot print at the crime scene, and walnut dust, which is used to polish statues, suggesting that the assailant may have been a park employee. The problem lies in that the assailant is a nonsecretor, that is, the semen found on the victim contains no sperm, and therefore there is no DNA evidence. A suspect is found who works in the park, whose boots match the boot print at the crime scene, and who is a non-secretor. Regardless, in the dawning age of genetic justice, this is not enough to convict, in the fictive world of television at least. The link between suspect and victim must be directly seen through a genetic trace.

Contested ways of knowing are also highlighted in a stranger-than-fiction case in Canada. A Saskatchewan doctor, John Schneeberger, was accused of drugging and sexually assaulting two female patients. When the women complained to the police, inspectors took blood samples from the doctor in order to test his DNA against the DNA in the semen found in the victims. Three separate tests were performed, and none produced a match. No charges were laid. The testimony of the eyewitnesses, complicated by their drugging, was rejected in light of the seeming conclusiveness of the DNA tests. Yet subsequently, as a result of police vigilance, it was discovered that the doctor had implanted a tube in his arm, which held someone else's blood. Because he was a doctor and the police did not have the resources to hire a technician to draw his blood for the DNA test, the suspect had drawn the blood himself under their supervision. Finally, a forensic technician conducting the DNA testing noticed that the blood was old, as though it had been out of the body for a period of time. This observation led to the discovery of the implanted tube. A fourth DNA test using Schneeberger's actual blood confirmed that he was, in fact, guilty of the crimes. Again, however, this case demonstrates that other forms of knowing, including eyewitness testimony and physical evidence, are reduced to secondary status in the face of molecular representation – which carries absolute weight.

The very real implications for subjectivity of genetic knowing can be seen, for example, in the radical reduction in numbers of babies being born with Down syndrome read as an instance of the broader phenomenon of the quest for bodily perfection. The example illustrates that the tools through which we are pursuing this objective are now more transformative than cosmetic. The International Clearinghouse for Birth Defects Monitoring Systems – an organization that compiles data from nations around the world on the prevalence, screening, and treatment of birth defects – reports that during 2001 51.8 per cent of Down syndrome pregnancies that had been prenatally diagnosed were terminated. In the age category where the risk of the occurrence of Down syndrome in the fetus increases dramatically (thirty-five to thirty-nine years of age), termination rates were regularly above 80 and 90 per cent in Western European countries (International Centre on Birth Defects Monitoring Systems 2003, 16–18). Canadian data were collected from Alberta, where the termination rate was 30 per cent. Obviously termination rates are directly related to the ease of access to abortion and the general cultural attitudes toward it (Health Canada 2002, 4). Yet the data clearly indicate that pregnant women in the 'risky' age ranges are increasingly expected to engage in prenatal screening, given its very high accuracy rates. Further, upon detection of Down syndrome, the preferred option is to terminate the pregnancy (obviously, again, in certain countries). What these statistics demonstrate is that the shift towards termination over the last decade is related to the genetic quest for an improved, if not perfected, human, a subject already of the future before it is even born – a situation directly produced out of the epistemological power of genetic testing.

While each of these examples, fictional and non-fictional, is anchored in actual knowledge practices, they rest upon a larger set of genetic myths. In the *CSI* example, the production of knowledge depends upon a belief in the 'culture of the trace' (Joseph and Winter 1996). In other words, the human body always can be unproblematically regenerated from its smallest traces through technologies of surveillance and simulation. Both the *CSI* and Schneeberger examples illustrate the bolstering of the DNA mantra as truth. DNA is figured as a lens onto a fully objective set of facts that cannot be tainted because they transcend the messiness of human memory. In essence, they promote the belief that DNA tells the truth, while human beings can only tell stories. Finally, the reduction of Down syndrome in the Western world augurs the emergence of the genetic super-human. The genetically enhanced – or at least corrected – human being is a staple of scientific representations from B films to biotech advertising. Yet 'he' is also the gold standard without referent to which van Dijck refers.

These myths are some of the social science fictions underlying both actual and projected scientific developments. As noted above, social science fictions are narratives that reassert the centrality and importance of applied science and the scientific project. Whether for good or for evil, they re-inject a sense of wonder into encounters with science. Working as epistemological frames, they produce a number of knowledge effects, including the conflation of future possibility with present reality, a conflation that results in new forms of scientific determinism. If the future is already now, then the political question becomes how to address its operations, not how to choose between future possibilities. Correspondingly, the history of the scientific enterprise is obscured and even effaced; historical knowledge is disprivileged as a site of productive critique and analysis. Social science fictions also re-mystify science, ensuring its unchallenged status as a knowledge form, and they have been used by both advocates and critics of biotechnological practice. Favoured sites of knowledge production expand and become more complex – one learns about science not only from scientists and governments, but also from fiction, film, television, and other cultural forms.

The field of biotechnology has been influenced by social science fictions since its beginnings. These fictions do very particular work in relation to the epistemology of visioning, for they bridge the conceptual and imaginative gap between the ability to see or perceive the self as a molecular being, and render visible the promises of biotechnology, which are not yet accessible through scientific practice. Consequently, they serve as a vector for both scientific and public agency. For scientists, social science fictions empower and protect their claims, their expertise, and their social function. At the same time, for the public, social science fictions translate otherwise inaccessible knowledge into a set of social ramifications that can be recognized and negotiated. These processes are both simultaneous and contradictory. They also produce uncertain knowledge. The maps that biotechnologies generate are at best unpredictable, non-linear trajectories, based as they are on the fragments of the body (Kellner and Best 2001, 110). Members of the public are left with questions about the advance of scientific knowledge, and its current status: are there laboratories presently attempting to clone a human being? Conflicting information circulates daily. Is the scientific knowledge available to pre-select for sex in in vitro fertilization techniques? In early 2005, scientists once again announced the discovery of the 'gay gene.' Throughout the 2000s, various rogue scientists claimed to have cloned a human being. In mid-2007, Dr Craig Venter (one of the leading scientists in the private-sector effort to map the human genome) claimed to have the capacity to build a new life form, not through manipulating the genetic material of existing organisms, but by building one 'from the ground up.' We've heard all of these claims before. Are they true this time? Are these social science fictions? Or are they science fact? And, more importantly, how does one know? How does one discern? The field is rendered muddy by the dominance of social science fictions. Biotechnological knowledge is uncertain knowledge.

Becoming Biosubjects: New Modes of Biosubjectivity

While most critics of biotechnology seek a means through which to inject social concerns into the realm of technoscience, the issue is more, as legal scholar Alain Pottage suggests, 'how to construct a model of individuality that takes into account the radical contingency of the bond between the social and the biological' (1998, 168). This observation is echoed by Novas and Rose's claims for somatic individuality, which represents a 'mutation in "personhood"' (2000, 485). They argue that new and direct relations are being established between self and body, registering in practices as diverse as bodily modification and theoretical corporealism. While there are a number of tensions or shortcomings in

Novas and Rose's notion of somatic individuality – the absence of law in its constitution, the focus on the material body to the exclusion of the informated body, and the totalizing treatment of the body – we agree with those authors, as well as with Pottage, that one's understanding of the individual is at stake in the biotechnological realm. More fundamentally, subjectivity is in flux. Although the individual is the social actor bearing legal and political rights, the subject is the accumulation of subjective experience; it comprises the ways in which the self is inhabited. Scientists and governments appeal to the individual in acquiring acceptance for biotechnological therapies that directly intervene on the body. They claim an individual's 'right' to know, and to make 'personal' decisions about his or her health, reproductive choices, and other life strategies, as if such knowledge and its consequences existed in a social vacuum where only the 'truths' of DNA matter. In this book we prefer to think conjuncturally and historically about what these technologies mean as part of new ontologies and epistemologies of the body. While the following analyses are concerned with the functioning of notions of individuality, they primarily explore new conceptual understandings of the self and subjectivity, which are structured by, and surface during, one's movement in and through biotechnological social relations.

In the chapters subsequent, we trace the emergence of what we see as a new mode of subjectivity – the biosubject. Although intimately connected to the body, we will argue the biosubject can no longer draw solace from an immutable, easily recognized, bounded body. As we shall see, it is constituted in a web of biotechnologically influenced social practices. Yet the biosubjects we examine are not exclusively human; they are simultaneously objects and subjects of institutional operations, and they mark both enhanced and limited forms of social agency. In that sense, then, the following arguments are in agreement with Susan Squier's contention that the biosubject exists in a kind of 'paradoxical space,' in which purity abuts hybridity and calls into question the very notion of human-ness (1998, 376). Both patients and disease vectors emerge in genetic disease–control protocols as biosubjects that require control and management; transgenic higher life forms of all sorts (from plants to humans) become subjects at law and objects of patents; fetuses and mothers become biosubjects in conflict with and established by embryonic technologies; and marginal populations are redefined as risky subjects and are objectified through the same DNA sampling techniques.

We argue that the biosubject is multiple – it is multiply determined and acts as a site for varying subjectivities. It is differently mobile, no longer marked solely by external factors such as race, class, gender, or geography – it is a subject whose ability to move within social institutions and social roles is determined by the invisible attributes of identity, yet is visible only through biotechniques. It is a subject that, we suggest, can no longer cling to nature as a cornerstone of identity, but rather is thoroughly artificial and prefers it so. The biosubject is desexualized – sexual relevance is supplanted by a set of biotechnological practices that are distinct from it, merely requiring a contribution of genetic fragments. Our case studies will demonstrate how biosubjects are caught up in new regulatory techniques, which not only recommend or require certain behaviours, but prescribe the ways in which one is expected to be, at an ontological level. One is supposed to be pure, efficient, and productive, and there are genetic techniques that can be used to assist in these goals if people are unable to achieve them on their own. As individuals, then, biosubjects are responsible for managing their selves, including their own messy bodies, their economic value, their impacts on, or risks to, other subjects, and their implications for future subjects. The biosubject is, therefore, we claim, a social science fiction in its own right.

The following four chapters explore emergent biosubjectivities in various sites of governance in Canada. Chapter 2, 'DNA Identification and Genetic Justice,' addresses the expanding phenomenon of genetic surveillance, using the criminal justice system as case study. Examining two different sites of cultural production, it analyses how certain popular and public narratives about DNA and crime act, alongside regulatory practices, to legitimize genetic surveillance by state agencies. One such site is the 'CSI effect' – the impact that popular forensic television programs have on criminal investigations and on public expectations of the criminal justice system.

As will be seen in chapters 3 and 5 as well, the media emerge as one of the central mechanisms through which citizens learn about biotechnology and, therefore, their frames are key to public understanding. Chapter 2 explores this phenomenon through the cultural events of popular media trials, concentrating on the case of David Milgaard and his struggle to be exonerated for a 1969 murder in Saskatoon. Tracing the elements of the media narrative involved in constructing the meanings of this case, it becomes clear that it is another rupture point between criminal justice experts, on the one hand, and media experts and the public, on the other.

Both of these instances demonstrate that cultural understandings of biotechnology significantly affect institutional processes and organizations authorized to manage biotechnology. This disrupts the frequent assumption that the flow of influence is always from institutions to the public domain. When combined with the actual Canadian legislation that establishes DNA warrants and a DNA databank, the effect of these public narratives becomes apparent. The policy process deflects attempts to politicize genetic surveillance, evacuating the language of bodily integrity rights from the legislation and replacing them with privacy rights. Consequently, a shift has occurred in the relationship between the citizen's body and that of the state. Privacy rights protect the information drawn from the body, not the body itself, thereby opening it up to genetic surveillance upon demand, as exemplified by the growing practice of DNA dragnets. As a biosubject, the citizen becomes responsibilized, with no moral ground upon which to refuse the molecular gaze, if the state can guarantee genetic privacy. The result is a loss of bodily sovereignty on the part of the citizen and an increase in the power of the state justified through the social science fiction of a perfected genetic surveillance system whose logical endpoint is a crime-free society.

Chapter 3, 'The Sexual Politics of Biotechnology,' maps and analyses the shifts in the Canadian governance of reproduction in a biotechnologized era. From the 1960s to the present, we witness a dramatic shift in governmental focus and rationale – from the regulation of conception and the recognition of women's bodies and their rights as subjects, to the regulation of fertility and the resulting effacement of the woman. We read this as a shift from discipline to biopower. Embryos and families thus emerge as the biogovernmental objects and subjects of choice, not the individual subjects who make them. Unlike the domain of patent law, the legislative branch did eventually act, but, as in patent law, the courts played a significant role in this process. Dramatically unlike both the DNA databanking regime and the legal disputes about the patenting of higher life forms, where the public and the media were largely silent, in the domain of reproduction, Parliament was forced to engage in the biogovernmental tactic of politicization in order to quell conflict, opposition, and even scandal. The shift from conception to fertility can thus be understood as a long and difficult – but ultimately successful – attempt by the state to shift public attention away from women wanting to prevent pregnancy, to heterosexual couples wanting children.

The analysis draws out four biogovernmental events across Canada's recent reproductive history. We first consider the challenge to Canada's abortion laws over the 1970s and 1980s by Henry Morgentaler. Second, we take up the events of the summer of 1989, when, in three separate cases in Canada, potential fathers were using the courts to prevent their former partners from having abortions. Third, we consider the scandal-plagued Royal Commission on New Reproductive Technologies and the scientific, public and political reaction to its work and recommendations. Finally, we follow the ten-year path from the Royal Commission to Canada's eventual legislation on reproductive biotechnologies, *The Assisted Human Reproduction Act.* These events combine to mark a process by which the state shifted its governmental object, new biosubjects emerged, and new sexual politics became necessary.

Chapter 4, 'Biopatents and the Ownership of Life,' considers the development of the biosubject in and through law. Over the past thirty years, patent law and biotechnology have commingled in attempts to patent altered life forms – a practice known as 'biopatenting.' Less controversial when focused on bacteria and plants, biopatenting became the subject of heated political and social debate around the world when American scientists attempted to patent a mammal, specifically a mouse transgenically altered to produce cancerous tumours. The Canadian courts broke ranks with the rest of the Western world and the Canadian government's own advisory body in holding that the Oncomouse was not a human invention, and, therefore, not patentable. This chapter examines the talk about a transgenic mouse – the Oncomouse – as it pursues its legal-administrative journey all the way to the Supreme Court of Canada.

As is the case in the area of criminal justice, public dissent is highly muted in the Oncomouse case, and the state has remained completely inactive in the area of biopatents. The concurrent absence of media commentary has meant a marginalization of the public. These ongoing silences have empowered the legal system to define the elements and ramifications of the biosubject, making it a biojuridical subject in this instance. Yet the frames set by the courts and policymakers have impacts that exceed the administrative domain and shape the broader cultural imaginary. The biojuridical subject emerging in biopatenting disputes is a subject amenable to property status. Echoing the microscopic biosubjects of chapter 5, the biojuridical subject of intellectual property law is not necessarily human.

Bioterrorism and epidemics have come to inhabit a central place in the popular imagination, especially since 9/11, and constitute the focus of chapter 5. However, bioterrorism in particular remains a social science fiction in the sense that today it is treated as a security priority, despite the fact that there has been only one occurrence – the 'Amerithrax' case – that followed the 9/11 attack on the World Trade Center.

As a result of the threat of bioterrorism and global epidemics, a new biosecurity system is under development around the world, but it too is a social science fiction. It exists largely in computer simulations. Nevertheless, as this chapter argues, the emerging biosecurity regime is having a significant biogovernmental impact by establishing a set of preparedness measures and bureaucratic reorganizations that have as their goal the sealing of leaky boundaries and the control of bioagency. Following the emerging international standards for biosecurity has become a new basis for establishing a hierarchy of nations – those that are responsibilized into controlling bioagency and those that are the source of dangerous bioagents. However, bioagency is very difficult to control and borders are always leaky to some extent.

In tracing these developments, this chapter examines dominant definitions of bioterrorism and security to see how they have shaped the nature of the biosecurity system aimed primarily at bioterrorism. In developing this system, authorities have found that truly international responsibilization strategies directed at two of the most important bioagents – scientists and governments – have failed. Therefore, individual nations have begun to develop their own biosecurity systems, according to a set of standards that have been formulated primarily by the United States. These systems tend to focus on vaccine research and stockpiling, research into microbe detection and defence technologies, and bureaucratic centralization of bioterrorism preparedness. These developments have become part of the larger strategy involved in the 'war on terror.'

We examine how Canada has reacted to these biosecurity requirements through its own vaccine stockpiling and microbe defence as well as the formation of the Department of Public Safety in 2003, which combines border services, intelligence services, corrections, parole, and the Royal Canadian Mounted Police under one ministry charged with providing national security. The Department of Public Safety, combined with the recently formed Public Health Agency, would provide for Canada's security in a bioterror attack or an epidemic. The Toronto SARS outbreak of 2003 is analysed as a case study of the biosecurity system in action and the pressures faced by governments and populations to seal leaky borders, even as the impossibility of complete control of bioagency becomes apparent.

Taken together, these case studies of criminality, replication, property, and terror demonstrate the ways in which popular, legal, administrative, political, and cultural narratives intertwine. They map a range of avenues through which the shift from bodily integrity to contingency is produced.

Each case represents a different point on a continuum of biogovern-mental action, involving different relations and roles for the govern-ment, courts, interest groups, citizens, and the media. We recognize some of the broader themes and figures that are consistent across a wide range of different cultural and political locations as Canadians make sense of biotechnologies. And finally, in each case study we see a differ-ent face of the biosubject.

Our notion of biosubjectivity attempts to reintegrate the body, in its contingent manifestations, with its biotechnological mediations. This oc-curs not only in the lab or the doctor's office, but also in our legislative, criminal, civil, and popular arenas, creating new public cultural spaces wherein one's body and one's identities are up for grabs. Understanding the intricate and often obfuscating ways in which beings are implicated as biosubjects means coming to terms with new discursive and social frames for living, placing contemporary examples into larger historical, legal, and social contexts, and exploring critically how narratives are con-structed to explain these events in the media and state discourses alike. *Becoming Biosubjects* suggests ways to step outside of both the dystopian nightmares of science run amok, based, as they often are, on a fear of the future, and the romantic, even nostalgic, futurism of pro-biotechnology arguments, in order to view them more accurately as structuring social science fictions. Indeed, this book is grounded in the material and dis-cursive conditions of biotechnological production in Canada, in order to reveal specific political, economic, social, and cultural strategies for their integration into everyday life and consciousness. In this sense, then, *Becoming Biosubjects* offers a way to rethink biotechnologies as a persistent and compelling reality, a reality that everyone must confront, process, and negotiate, if we are to function, now, in the brave new biotechno-logical world.

CHAPTER TWO

DNA Identification and Genetic Justice

GREG. Look, I don't have time for your humour ... Warwick tells me his home invasion is my top priority and I'm still backed up on Catherine's no-suspect rape. I'm serving many masters, you know what I'm saying?
GRISSOM. Greg, this is your DNA lab. You are the master. We serve you.
GREG. Well, your stuff just moved to the top of the pile. What've you got?
GRISSOM. Individual DNA from a mob attack on a cabbie, and the cabbie's clothes.
GREG. Get ready to match the stars.

CSI: Crime Scene Investigation (season 3, episode 9, 'Blood Lust')

A man lies on a wheeled hospital bed in a sterile room. The only other occupants are a medical doctor and a prison guard. Attached to his arm is an intravenous bag containing chemicals that will stop his heart when released into his bloodstream. A curtain is pulled back from the room's only window, and behind it are seated witnesses to the execution that is about to take place. The man is a convicted murderer in the state of Nevada, and he is about to undergo the enactment of his death sentence. At the last moment, however, the state governor issues a reprieve; another crime with the same characteristics has been committed while the prisoner was incarcerated, suggesting that he may not be the murderer after all. Another round of investigation must now begin to determine whether the condemned man is guilty, but this time it will be different. A new tool has been added to the police repertoire – DNA analysis.

This is an opening scene from an episode of *CSI: Crime Scene Investigation*. It is a disturbing scene, evoking both a sense of biblical justice for a cold-hearted killer, and, at the same time, a twentieth-century-inspired fear about the oppressive potential of a close relationship between science, medicine, and police. The tension between justice and fear characterizes our current relationship with biotechnologies, which allow authorities to match DNA traces to specific individuals. What do citizens give up as they empower the state to place their genetic codes under surveillance? Although DNA-matching technology is relatively new, it has been in place long enough for people to begin to perceive the outlines of an emerging genetic surveillance system. As David Lyon (2001) argues, 'In a world of identity politics and risk management, surveillance is turning decisively to the body as a document for identification, and as a source of prediction' (72). However, if bodily surveillance at the molecular level is a characteristic of late modernity, it is hampered by legal techniques and cultural values from an earlier modernity – techniques and values that protect the body from access upon demand by the state or anyone else. In this situation, the biogovernmental problematic becomes a question of how to overcome those restrictions in culture and in law. This chapter argues that the process of implementing a genetic surveillance system in criminal justice has been, in part, a process of changing the law, as well as an appeal to popular culture to ease the entry of this new surveillance technology into society. As a consequence, the relationship between the citizen's body and the state has changed.

Genetic Surveillance

Biotechnologies of surveillance enter into our society through biogovernmental conditions that reflect privatization, politicization, objectification, normalization, and responsibilization. Yet how these processes work themselves out in connection with genetic surveillance are different from how they operate with other biotechnologies. Over the past few decades, social scientists have paid increased attention to questions of surveillance, and its converse, privacy, as a result of the deployment of new electronic surveillance technologies. Initially, these technologies were used in the workplace, but spread to encompass more and more spheres of everyday life. As Anthony Giddens (1985) points out, surveillance has long been one of the central institutions of modernity, and new surveillance technologies enter into society through pre-existing relationships between citizens and surveillance. However, electronic surveillance and surveillance of human genetic material involve overturning some of these pre-existing relationships and redefining them. In that sense, their effects are revolutionary.

In the social sciences, analysis of surveillance has traditionally followed either Marxist or Weberian lines, viewing the logic of surveillance as motivated by the needs of capitalism to promote ever-greater degrees of efficiency and productivity in the working class, or by the need to further enhance the instrumental rationalization that underlies modern institutions. Beginning in the 1980s, Foucault's argument that power is a ubiquitous social field that permeates all relationships began to revolutionize the field of surveillance studies. As one element of the social field of power, surveillance has the effect of bringing people under stricter regimes of self- and social regulation through practices that render populations and bodies observable: it engenders a surveillance society.

Foucault's analysis was directed at practices of early modernity, but subsequent scholars have found his language useful for understanding the highly technologized surveillance of the late twentieth and early twenty-first centuries. Robert Castel (1991), for example, suggests that today virtual selves develop through expert coding and are circulated and managed independently of the bodies to which they refer. Mark Poster (1996) goes even further, pointing out that the databases formed through contemporary monitoring produce a 'superpanopticon,' which constitutes subjects through textual production (182). He argues that the database has certain new properties that make it particularly effective. It is electronic and digital, and therefore easily transferable in time and space. It has no particular identifiable authors, but is authored by anyone who enters data, including machines that automatically record information. Nevertheless, it belongs to someone – the corporation, the government, the police, the military, the hospital, the university – and amplifies the power of its owner (183). Furthermore, everyone knows about these databases, and there is a curious mixture of unease and acceptance of their operation.

Shoshana Zuboff (1988) refers to this form of surveillance as 'informating' – that is, the use of information technology to automatically record transactions in order to create an electronic information environment. In this way, decision makers have an immediately visible window on the processes of workplaces, marketplaces, and public places. They know, in real time, how productive people are, what they are purchasing, and where they have been. Informating describes a more generalized process of the superpanopticon. The one who is under surveillance provides the necessary information. There is no need for a carefully designed architecture of power based on complex bureaucratic organization operated by experts on security, labour, and consumption.

Voluntary transactions are automatically recorded and converted into texts through the cables and circuits that form the pathways of power and knowledge. These texts comprise the discourse of surveillance; they constitute their own object by producing a collective account of a multitude of private actions.

Synthesizing many of these notions, William Bogard argues that, in its current form, surveillance forms part of an 'imaginary of control' (1996, 9). Surveillance is more than a tool in the maintenance of social order; it is also a fantasy of power, which, in post-industrial information societies, takes the form of codes that simulate a perfect surveillance. In other words, surveillance is increasingly a social science fiction, another form of imaginary, in which, at the push of a button, anything can be made visible and knowable. Consequently, the problem addressed by surveillance shifts from a problem of control to a twofold problem of how to code information and enter it into databases, and then how to cut the time of transmission of information to zero, thereby closing the gap between the actual and the virtual – the body and the code. From this perspective, people are not under surveillance, but rather coded information about them is collected. As a result, the struggle between control and resistance becomes less important than a logic of virtualization. Resistance does not overcome surveillance; it simply focuses the gaze.

Concurrent with these developments in surveillance logic, and as we outlined in the introduction, a new conceptualization of the gene has emerged over the past two decades. Before genes can be manipulated, broken apart, analysed, and converted into information about identity, they must first be imagined. They must be represented as something that is open to human observation and intervention. This is necessary for public legitimation of genetic research and is an essential component of the biotechnological imaginary that is forming in Western culture. Van Dijck argues that the gene has become an interactional information map, translated into digital code and rendered readable as an unambiguous text. When combined with the logic of simulated surveillance, genetic surveillance involves informating and simulating, both of which work to bridge the distinction between the body and the code. This is how the biogovernmental process of objectification operates in this particular biotechnology. The imaginary of control turns its gaze on the gene as an objectified information map, which identifies the person and links him or her to places and events with a certainty lacking in other forms of biometric surveillance. What is under surveillance, however, is not the person but the digitally recorded DNA code, a code that is periodically subjected to observation in matches to crime scene DNA codes. This simulated mode of surveillance means that as soon as crime scene DNA is coded it can be matched immediately to an offender's DNA code (if it has been previously recorded) in a much shorter period than would be the case in more traditional modes of investigation. As a biogovernmental process, genetic surveillance involves objectification of the gene through the language of information science, translating it into a form of information subject to the same kinds of manipulations as other forms of information.

Bodily Integrity and the Law

From the perspective of law enforcement, the technology and the logic of genetic surveillance are in place, but there are roadblocks to its full implementation. The problem lies in gaining legal access to the DNA codes of citizens and culturally legitimating that power of access without initiating a public backlash. It is a problem that emerges from the intersection of law and culture, at the heart of which is the notion of 'bodily integrity.' The concept of bodily integrity is recognized in the common law and is generally seen as implicit in section 7 of the Canadian *Charter of Rights and Freedoms*, which states, 'Everyone has the right to life, liberty and security of the person and the right not to be deprived thereof except in accordance with the principles of fundamental justice.' Arguably, section 8 of the *Charter* also has application: 'Everyone has the right to be secure from unreasonable search or seizure.' Generally speaking, bodily integrity entails sovereignty over one's own body and the right to make choices about invasive entries into the body. In a cultural sense, as discussed in chapter 1, bodily integrity is the primary way of thinking about the relationship between the body and the self within modernity. It is based on an anatomical model of what the body is and provides a boundary between the self/citizen and the state. This is why the state cannot legally penetrate that boundary without an overriding community interest.

This principle has been the basis for several important Supreme Court decisions over the past two decades in Canada. For example, in the case of *R. v. Morgentaler* (1988), the Supreme Court struck down the *Criminal Code* provision that criminalized therapeutic abortions outside of an approved administrative regime, arguing that the provision clearly interfered with a woman's bodily integrity, and therefore violated section 7 of the *Charter*. In the case of *Rodriguez v. British Columbia (Attorney General)* (1993), Sue Rodriguez, a terminally ill patient, challenged the *Criminal*

Code provision that criminalized assisted suicide, arguing that it violated her bodily integrity. The Supreme Court agreed but went on to point out that bodily integrity is outweighed by the principle of the sanctity of life and therefore is not an absolute principle.

Another relevant area of common law in which bodily integrity is invoked lies in police seizure of medical samples. This situation occurs most commonly in impaired driving cases, where an injured suspect has given blood and urine samples to medical personnel. These samples are a source of evidence about blood alcohol levels. However, in the case of *R. v. Dyment* (1988), the Supreme Court held that in the absence of a warrant the seizure of a suspect's medical sample is a 'gross violation of the sanctity, integrity and privacy of the appellant's bodily substances and medical records' (263). In *Dyment* and other cases, the court has used bodily integrity to apply not only to the physical body but also to information about the body.

This was the situation when DNA-matching technology entered the Canadian criminal justice system. The actual regulation and use of DNA identification technology in criminal justice dates back to 1986 England. In 1985, Alec Jeffreys and a team of geneticists at the University of Leicester publicly announced a technique for linking DNA samples to specific individuals. Right at that time, a murder investigation was ongoing in the Leicester region, and police immediately seized upon Jeffreys' discovery to aid them in their investigation. Two young women had been sexually assaulted and murdered, and the police had few clues about the identity of the attacker, other than semen found on the victims. Armed with the new technology of DNA identification, police requested that all men of a certain age range, in the three surrounding communities, submit to DNA testing; however, no matches could be found to the crime scene DNA. The case was resolved when someone overheard one man telling others in a pub that he had been paid to stand in for another man during the mass DNA testing. After questioning the witness from the pub, the police learned the identity of the man who had eluded the DNA net – Colin Pitchfork – who was apprehended, tested for a DNA match, and became the first person in the annals of criminal justice to be convicted through DNA matching.

The technology spread quickly, making its first appearance in Canada in 1988.[1] In many ways, DNA matching was a revolutionary shift in criminal identification. Just at a time when the Supreme Court was strongly defending the notion of bodily integrity, a promising surveillance technology appeared that would require easy access to citizens' bodies in order to operate at its greatest efficiency. Consequently, the courts were uncertain how to treat this new technology. Some excluded DNA evidence as a violation of the *Charter*, while others accepted it. Police found themselves rummaging through garbage cans and ashtrays looking for DNA samples because there was no common law or statutory authority to seize samples directly from suspects, a situation confirmed and perpetuated by the first Supreme Court decision on the matter in 1994, in the case of *R. v. Borden*.[2]

That decision indicated to the government and law enforcement that the common law doctrine of bodily integrity posed too great a barrier to implementation of a genetic surveillance system. In response, the government moved quickly to legislate the use of DNA identification technology in criminal justice. After public consultation in 1994, the *Criminal Code* was amended in 1995 to include provisions for DNA warrants in the investigation of designated offences. Another public consultation occurred in 1996, leading to the passage of the *DNA Identification Act* in late 1998. Through these acts, police were empowered to seize DNA samples from suspects, and, upon conviction for certain types of offences, store DNA samples in a DNA databank, which went into operation in July 2000. A number of inevitable *Charter of Rights* challenges followed, based primarily on sections 7 and 8. However, the courts upheld the new laws as justifiable in a free and democratic society.

The sampling procedures allowed by the legislation include plucking hair roots; taking buccal swabs of the lips, tongue, and inside cheeks; and using a lancet to take a blood sample. Samples can be used only for the strict purpose of forensic DNA analysis in the investigation of a particular case, and there are strict punishments for those who violate that restriction. Police officers executing DNA warrants and laboratory personnel conducting DNA analysis must ensure that precautions are taken to respect the privacy of the person. If a suspect is cleared by DNA testing or is acquitted of the offence, his or her DNA samples must be destroyed. But there are circumstances wherein the judge can defer the destruction of the sample, and the information obtained from it, for any period he or she considers appropriate. Generally, this is meant to permit the police to investigate the same suspect for other designated offences, in this way broadening the scope of the DNA warrant beyond the immediate offence.

There are certain absences in the legislation that are as striking as what is present. Since the issue is never mentioned, the act implies that there is no need for a warrant if a person consents to the removal of bodily substances. However, the act does not define what constitutes proper consent. The warrant itself can be executed without any requirement of consent on the part of the suspect, and, according to section 487.07(1)(e), an officer may use 'as much force as is necessary for the purpose of executing the warrant.' A final point of note is the list of designated offences. They range from acts of piracy and hijacking to violent and sexual offences, to arson and breaking and entering. Attempts to commit these offences are also designated. Although the list is generally limited to those offences in which bodily substances may be expected to be left at crime scenes, it is considerably more extensive than that recommended by the privacy commissioner.

The DNA Identification Act established the National DNA Data Bank in Ottawa under the jurisdiction of the RCMP. Safeguards in its operation were included to protect the privacy of individuals. Only authorized law-enforcement personnel are empowered to inquire into whether or not a suspect's DNA profile is within the databank. The RCMP commissioner could communicate that information to foreign law-enforcement agencies and allow access to anyone considered appropriate for the operation of the databank or for training. Criminal penalties would apply to anyone who violated the conditions of privacy contained in the act. In addition to information profiles, the DNA databank would also store bodily substances for future analysis. There are two indexes in the databank – a convicted offenders index and a crime scene index of DNA samples. Generally, the information and the biological samples would be stored indefinitely, although the information would be destroyed in cases where the offender was acquitted. In the case of young offenders, the information would be destroyed after ten years, five years, or three years, depending on the severity of the offence.

There are two types of designated offences: primary and secondary. Primary designated offences are generally the most serious violent and sexual offences, in which DNA evidence is likely to be useful in investigating the crimes. There is a presumption that an offender convicted of a primary designated offence will be required to submit a sample to the National DNA Data Bank unless countervailing substantial harm can be demonstrated. Secondary offences are serious – but less serious – offences for which the Crown must apply to the court for retention of a sample from a convicted offender. The court has more discretion in determining whether or not DNA sampling is appropriate in these cases. The legislation is retroactive; people convicted of designated offences before the passage of the act may be sampled if they have been declared 'dangerous offenders,' if they have been convicted of more than one sexual offence, or if they have been convicted of more than one murder.

In 1999, a further addition was made to the DNA databank provisions through Bill S-10, a Senate bill, which adds certain military offences to the list of designated offences and provides for a Senate review of the legislation after five years. Overall, the provisions of the databank and the warrant were products of a series of debates and fears expressed in policymaking and were carefully crafted to survive the inevitable *Charter* challenges they were soon to face. Contestations and rupture points among major policy stakeholders, prior to the passage of the legislation, uncovered debate and competing assumptions about the meanings of genetics and its intersection with crime control, meanings that were articulated in terms of individual rights versus public safety.

Entry of genetic surveillance technologies into criminal justice involves complex changes in the powers of authorities and the relationship between citizens and the state. Framing the legislation requires care in presenting the issue to the public for fear of backlash against expanded state power. Yet there has been little cause for concern in most countries where the technology has been implemented. Public debate and interest has been absent in the implementation of genetic surveillance – partly as a result of the interrelationship between high-level fears of crime, a growing comfort with the electronic surveillance systems that have become an infrastructural element of everyday life, and cultural fascination with genetic science (see Gerlach 2004). In this chapter, we explore two cultural sites where these elements come together to produce representations that place a positive spin on genetic surveillance, a spin that facilitates and legitimates the elimination of bodily integrity as a value in legal culture: the 'CSI effect' and media coverage of sensational trials.

The Real 'CSI Effect'

Premiering in October 2000, *CSI* and a host of television sequels and imitators have popularized the science of forensic investigation to an unprecedented degree. Such popularity is a product of a number of factors, chief among them the enticing format of the program. Forensic investigators pursue, discover, and analyse the minute traces left by human bodies at crime scenes, appealing to those who love scientific puzzles. The show is highly didactic, taking pains to educate the viewer in the science of the body, crime, and other miscellaneous facts about nature. For the most part, the result of the investigation is proven beyond question, by science rather than by law. The outcome is trustworthy not only because of scientific proof, but also because of the dispassionate nature of the investigators, who generally remain aloof from the people around them. These investigators provide a relief from the all-too-human cops, struggling with corruption, burnout, and personal motivations, who have characterized television police dramas for a number of years.[3] As one journalist stated, 'Clearly, the brilliant, slightly geeky investigator [Gil Grissom, the lead character in *CSI*] – with absurd amounts of knowledge on criminal profiling, blood spatter patterns, the rate insects eat flesh and other forensic sciences – has pushed aside the robust cops driving muscle cars who rough up pimps and druggies' (Humphreys 2003).

A significant amount of media attention, particularly in the United States, has been focused on what has come to be called the 'CSI effect.' Generally speaking, this effect is 'a phenomenon in which actual investigations are driven by the expectations of the millions of people who watch fake whodunits on TV' (Hempel 2003). Such expectations centre on the speed and almost miraculous abilities of forensics and biotechnologies to identify criminals; however, the phenomenon has a number of facets that make it an interesting sociological issue. It marks the entry into popular culture of what Joseph and Winter (1996) refer to as the culture of traces – the assumption that the body is constantly shedding clues that can be linked back to it by science. This is not new in popular culture and has been a staple of detective stories and police procedurals at least since Sherlock Holmes first picked up his magnifying glass. But *CSI* adds the power of new biotechnologies to the mix.

The real revolution in linking bodily traces to criminal activity occurred with the development of fingerprinting, which the British adopted from India in the late nineteenth century. In Canada, the first permanent fingerprinting bureau was established in 1910 by the Dominion Police, with Edward Foster at its head.[4] Other biometric measures have developed since that time, including measurement of iris patterns in the eye, face recognition technology, voice recognition, and hand printing. All of these measures are threatened with obsolescence by the onset of DNA identification. Although often referred to as 'DNA fingerprinting,' the new technology is not simply another form of biometric identification; it does not simply contain a trace of the individual. Rather, it contains a person's essence – a potential map of a person's fundamental characteristics – and much of the public debate about its use has centred on the resulting privacy concerns raised by the implications of genetic surveillance. Best-selling crime fiction writer Patricia

Cornwell noted that, in 1988, when she sent her first novel featuring a forensic detective, *Postmortem,* to publishers, it was rejected by seven major publishing houses because it was considered boring (Lipson 2004). Yet, by the time Kathy Reichs wrote *Déjà Dead,* another popular forensic detection novel, in 1996, publishers were eager to embrace it. What had changed in the intervening years was a growing public awareness of forensic detection, particularly DNA identification. The media narrative around the CSI effect draws on a list of preceding influences that have been instrumental in producing the current fascination with forensics, including the popular 1970s television show *Quincy,* about a medical examiner / crime investigator; medical-examiner fiction such as that of Patricia Cornwell and Kathy Reichs; the O.J. Simpson trial and the central role that DNA evidence played there; the Discovery Channel's popular 1996 television show *The New Detectives,* which re-enacted actual forensic detection cases; and finally, the 11 September attacks in New York and the coverage of the role that DNA identification played in the body recovery operations that followed. In addition, there has been intense coverage of local cases involving DNA evidence, including Guy Paul Morin and David Milgaard, in Canada. Morin was charged first with murder and then, after he was acquitted, with the sexual assault of his nine-year old neighbour, Christine Jessop. Although DNA testing in 1995 cleared him of the crime, an inquiry also noted irregularities by the Ontario Centre of Forensic Sciences at the outset of the case. Milgaard is probably the most celebrated figure of all the known miscarriages of justice in Canada, and his story will be outlined in greater detail later in this chapter.

The most publicly discussed element of the CSI effect is its impact on juries and legal practitioners. According to media reports, prosecutors are experiencing a problem with the CSI effect because juries expect too much of police investigators. They want to see DNA and other types of forensic evidence linking the suspect to the crime before they are willing to convict. One of the first cases to evoke the spectre of the CSI effect occurred in 2003. Robert Durst, a millionaire real estate heir in Texas, was accused of murdering and dismembering his neighbour, Morris Black. Black's head was never found, and Durst's lawyer exploited this gap in the evidence to argue that wounds on the head would have supported Durst's claim that he had acted in self-defence. The jury wanted to see forensic evidence before it would convict, and Durst was found not guilty. In 2004, in Peoria, Illinois, a gang member was accused of raping a teenager in a park. The prosecutor emphasized the fact that DNA from saliva found on the victim's breast matched that of the defendant. He was found not guilty. Some members of the jury felt that the police should have tested soil found on the victim to see if it matched soil found in the park. They had seen it done on *CSI*. Perhaps the most famous example of the CSI effect has been the trial of actor Robert Blake, in 2005. Despite considerable non-forensic evidence concerning motive and intent, jurors found Blake not guilty of murdering his wife because of an absence of gunpowder residue on his clothing. The type of gun used in the murder usually leaves residue.

In response to pressure for forensic evidence, prosecutors and defence lawyers are developing a number of strategies. Several states allow lawyers and jury consultants to ask prospective jurors about their familiarity with the *CSI* programs. This may have been the secret of Robert Durst's acquittal. His jury consultant, Robert Hirschhorn, worked to find jurors who were familiar with the programs (Willing 2004). In Arizona, Illinois, and California, prosecutors have begun to use 'negative evidence witnesses,' whose task it is to convince jurors that it is not unusual for forensic investigators to fail to find DNA and other forensic evidence at crime scenes (Willing 2004). Due to the development of a legal 'market' for forensic expertise, there is a growing phenomenon of self-appointed, underqualified forensic 'consultants,' with little training in the field, hiring themselves out to lawyers in order to add a scientific element to their case (Hempel 2003).

Although there is little empirical evidence that viewing forensic detective dramas on television has a direct impact on jury decision-making, there is evidence that there has been a rise in jury expectations and demands for scientific evidence at trial, which may be linked to a broader shift in popular conceptions of technology (Shelton, Kim, and Barak 2006) as well as a general decline in trust in criminal justice authorities (Tyler 2006). DNA evidence, in particular, appears to have a major impact on jury decision-making. Given that the odds of random matches of DNA can be as low as one in several billion, it is not surprising that juries and judges can be heavily influenced by DNA evidence – or its absence – in court.

Nevertheless, the presence of DNA evidence in a trial does not necessarily produce a sure outcome. Researchers have found that how DNA evidence is reported to jurors will have a significant impact. Jonathan Koehler, professor of behavioural decision-making at the University of Texas at Austin, and a member of the O.J. Simpson defence team, reported that the precise words used to describe the validity of DNA analysis can have a very significant impact on the jury's sense of guilt or innocence. In two experiments with college students, Koehler reported that he had asked the students to act as mock jurors to evaluate the efficacy of DNA evidence. In the first experiment, 249 students were divided into two groups. One group was told that there was only a 0.1 per cent chance that the accused would match the crime scene sample if he were not the perpetrator. The other half was told that there was a 1 in 1,000 chance that another person has the same genetic fingerprint. Of the students in the first group, 82 per cent found that the defendant was definitely the source, and 75 per cent concluded that the defendant should be found guilty. Of the students in the second group, only 43 per cent viewed the defendant as definitely the source, and 45 per cent concluded the defendant should be found guilty (Strauss 1996).

Clearly, Kohler's findings imply that presentation of DNA evidence has a large impact on a jury's perceptions. If jury members are told how unlikely it is that a suspect could match and still not be guilty, they tend to convict; the focus is on the defendant. If they are told how likely it is that someone else could also match, they tend to acquit; the focus is on those who are not suspects. Koehler points out that defence lawyers in the United States have been arguing for the 1 in 1,000 approach as the standard for presentation of evidence to ensure that the presumption of innocence is maintained. DNA evidence, therefore, is not simply a matter of the pure application of science; people's perceptions and misunderstandings of the actual meanings of the science come into play and must be accounted for in courts of law.

It is not only jury members who are influenced by representations of DNA evidence. Criminals are beginning to adapt to new, DNA-based forms of crime detection. American police have already reported a number of developments, such as rapists wearing masks, gloves, and condoms, and forcing victims to wash away DNA evidence; burglars wearing protective shoe covers; and prisoners caught tutoring one another on how to spread other people's blood and semen around crime scenes to trick forensic analysts (New Hampshire Police Standards and Training Council 2000). What these cases indicate is the degree to which DNA testing and banking have already become an integral part of crime management. They also suggest that criminals can and will develop tactics to counter the 'ultimate identifier' in the hopes of evading the DNA net. Some analysts fear that this development is also a part of the CSI effect.

As a consequence of the popularity of *CSI* and similar programs, colleges and universities in the United States are reporting a huge increase in student interest in forensic training programs. As CBS reported in 2004, 'Chemistry labs and criminal justice programs are what's cool on campus these days, as "CSI" – the nation's top-rated show last season – and its spinoff "CSI Miami" (which ranked 14th in the Nielsen ratings), have created a whirlwind of interest in forensic sciences' (CBSNEWS. com 2003). Unwilling to let such an opportunity pass them by, a number of post-secondary institutions in the United States have opened, or are planning to open, forensic science programs. Baylor University in Waco, Texas, reports a tenfold increase in enrolments since the program opened in 1999 (CBSNEWS.com 2003). Recently, the University of the Sciences in Philadelphia and Hilbert College in Hamburg, New York, have created forensic science minors, and the University of Baltimore and Saint Louis University have created majors in forensic sciences. New graduate programs have recently appeared at the University of California at Davis, Duquesne University, and University of North Texas Health Science Center. Others are considering new programs as well. According to admissions directors of these programs, there are long waiting lists, and, when applicants are asked why they are interested in forensic science, most of them say it is because of *CSI* (CBSNEWS.com 2003). Yet, according to one director of a forensic science program, despite the need for thousands more forensic technicians in U.S. crime labs, there is no job market for forensic practitioners because of restricted state budgets (CBSNEWS.com 2003).

This leads to the final element of the CSI effect – the impacts on actual forensic experts and their responses to expectations created by the television program. Much of the media discussion about the CSI effect has centred on the inaccuracies of the show and the false expectations those inaccuracies produce in the public consciousness. As one scientist stated, 'Nobody would watch a real show about forensic investigation because it's too boring. You'd be watching the back of my head as I look into a microscope for five hours' (Associated Press 2003). The ambivalence felt by forensic scientists toward the television phenomenon is encapsulated in the description of a session, entitled 'C.S.I. Effect,' held at the Fifth Annual National Conference on Science and the Law in Tampa Florida (14–17 March 2004):

Taking a look at television series that depict forensic scientists at work (e.g., Quincy, Profiler, X Files, Forensic Files, Medical Detectives to CSI), this panel will examine how the entertainment media influence jurors' expectations of the capabilities of forensic labs and the effects on undergraduate and graduate education (i.e. CSI 'scientists' also do investigations and law enforcement, there are unreasonable turn-around times for lab results, shows never follow cases through to trial, shows depict cross examination by the defense inaccurately, etc.). The panel will explore the following issues: have these popular television shows generated interest in forensic sciences and crime scene work among students? Is this interest reflected in increased interest in science, in general, among pre-college students? Have these shows created unrealistic expectations, in the courtroom and among laboratory personnel themselves, of what a forensic lab is, what it can do, and how it works? Have they skewed the public's perception of science in general?

In media interviews and profiles of individual investigators, forensic scientists admit to certain positive aspects of the CSI effect and the attention it focuses on their profession. Practitioners are gaining public recognition for the work they do, and report that they are in high demand to lecture about their work in schools and public venues. Some even characterize themselves as local celebrities. The television programs portray forensic investigators in a positive light – as specialized, highly trained experts who use science and interesting technology to assist in the capture of criminals.

Even so, forensic scientists interviewed in the press generally view the CSI effect as negative. It sensationalizes their work in ways that may prove disappointing to all of the students attracted to the profession by the glamour of the television representations. As one forensic investigator states, 'In most of the burglary and indoor homicide scenes I have worked, the house is so dirty I want to take a shower before I even begin processing. After several hours on the scene it looks like I have taken a bath in fingerprint powder, I am usually sweating from the Georgia heat, and even though the adrenaline is pumping, I am tired. Miraculously, the investigators on CSI, never have even a spot of fingerprint powder on themselves or their perfectly pressed clothing' (Shepard 2001). Indeed, most cases are not glamorous murder mysteries, but rather burglaries and stolen vehicles. Nor is it the job of crime scene analysts to prove who committed a crime; they simply attempt to reconstruct what happened at the crime scene. Generally, solving a crime relies on traditional forms of police work, with forensic evidence as a backup for other kinds of evidence. This is in contrast to television representations, where uniformed officers act primarily as bodyguards for scientific investigators, and CSI personnel regularly question suspects on their own while police detectives simply act as observers.

More importantly, practitioners are concerned about the unrealistic expectations that television dramas create among crime victims and their families, and among jurors and other legal practitioners. One spokesperson complained, 'Family members and others will call and want to know the results and why did it take four weeks? They did it in an hour on television' (*Vancouver Province* 2002). Unrealistic ideas about the time it takes to process forensic evidence is the primary complaint from the forensic community, which is tied into a number of other complaints about the lack of reality in television representations. Procedures such as DNA analysis take between twenty-four and forty-eight hours to complete; however, in real labs, they normally take two to four weeks because of the backlog of samples. Lack of laboratories, personnel, and equipment is largely to blame for the difference. Spokespersons also argue that, although much of the advanced technology used in *CSI* exists, real forensic labs cannot afford it on their restricted budgets and are also constrained by legal standards of accuracy. In an interesting attempt to directly affect real world forensics, William Petersen, who stars as *CSI*'s chief investigator, Gil Grissom, and who is also a co-executive producer of the program, testified before a U.S. congressional committee to argue for higher funding for crime-scene labs (Mason 2003).

In fact, when all elements of the CSI effect are taken together, there is really only one major concern expressed – the public misunderstanding of the actual practice of forensic investigation, and the exigencies under which it operates in terms of scientific and technological capabilities, legal constraints, and governmental and fiscal limitations. Media attention to the CSI effect is an exercise in promoting public understanding of forensic science. But why is this necessary?

Arguably, television programs such as *CSI* have tapped into an emerging cultural phenomenon, a scientific imaginary, and more specifically, a biotechnological imaginary, spurred on by the promises of DNA identification and other miraculous potentials that elevate genetic science to the level of a charismatic science reflecting future hopes. Given the place of forensic DNA identification technology within the biotechnological imaginary, the ways in which popular culture portrays forensic investigation have effects beyond generating misunderstandings of the science involved. Popular representations of DNA analysis in criminal justice are almost always positive, in the sense of rarely questioning the political power that lies behind the ability of state agents to enter into people's homes and bodies and seize biological samples. On *CSI*, investigators often request voluntary DNA samples, and sometimes, although rarely, people question their authority to do so. They are told it is their right to refuse, but if they do, they will simply be put in a cell until a DNA warrant is obtained. Viewers proceed through the investigation process from the point of view of the forensic investigators, for whom these political and ethical issues of civil rights are converted into depoliticized technical issues of legal and technological access. Assertions of bodily integrity rights in the face of state-sponsored genetic surveillance simply seem to be attempts to evade justice as embodied in scientific truth.

As scientists correctly point out, current television representations of forensic techniques are fictitious and inaccurate in several ways, but it does not matter. Forensic science and DNA analysis, in particular, have become signifiers in the popular imagination, with a signification different from that desired by scientific elites. The CSI effect marks a rupture point in the governance of genetic surveillance technologies. A contestation over the meaning of this technology is occurring as the standard scientific/legal narrative (preferred by scientific and legal elites) meets with a counter-narrative (emerging out of the public biotechnological imaginary). This hyper-real perspective on DNA forensic investigation, based upon media representations rather than laboratory practices, places extra demands, expectations, and responsibilities on law-enforcement personnel, since it does not account for the technological and institutional limitations of forensic practice. Regardless, attempts to educate the public about a more realistic understanding of the situation tend to fall on deaf ears, as more and more people tune in and sign up to train as forensic analysts.

The CSI effect is also problematic in its implied critique of previous modes of crime control. Arguably, one reason for the groundswell of public interest in DNA identification and other forms of forensic evidence is its perceived certainty, as opposed to the uncertainties and subjective nature of other forms of evidence and their treatment at trial. Popular portrayals of forensic science may feed, at least in part, into the high levels of distrust that currently inform public perception of North American criminal justice. For example, in 1999, in Canada, only 15 per cent of Canadians rated the criminal courts as performing well in helping victims; 33 per cent rated them as average; and 35 per cent perceived them as below average, with the remainder expressing no opinion (Statistics Canada 1999, 109). After a number of high-profile cases, in which innocent individuals have been imprisoned – sometimes for decades – until exonerated by DNA testing, there is considerable question about the efficacy of traditional forms of investigation and courtroom decision-making. In contrast, the Canadian public seems comfortable with current levels of policing, with 60 per cent stating that the police are doing well in enforcing laws, 29 per cent suggesting their performance is average, and only 5 per cent rating them as poor. Together, these court and policing statistics suggest that Canadians prefer a more punitive criminal justice system and see DNA evidence as an aid in ensuring more convictions through its objectifying qualities.

Ironically, the CSI effect, as an expression of popular support for a genetic surveillance system and a legitimation of scientized evidence-gathering in criminal justice, can undermine the ability of police and prosecutors to win convictions. The public wants convictions, but it wants certain convictions. It wants the biotechnology of crime detection to live up to its promise to produce clear pronouncements of objective truth. In some cases, this becomes more important than the actual conviction of defendants who seem guilty upon consideration of eyewitness's material and circumstantial evidence. If police have not conducted, or cannot conduct, a thorough forensic investigation, juries might not be convinced. This is a significant challenge to the authority of a long-standing and legitimated social consensus of what constitutes proper evidence. At the same time, the public seems impatient with the institutional and technological restrictions placed upon forensic science by legal and cultural traditions of bodily integrity. Apparently, that principle is expendable in the rush to allow the technology to fulfil its promise as the ultimate identifier.

Media Trials and DNA Evidence

The CSI effect is a relatively new phenomenon, but it draws upon older traditions of the scientific imaginary in popular and legal culture. Perhaps the most significant of these are well-publicized media trials that come to revolve around DNA evidence. In the 1990s, journalistic coverage of such trials spotlighted DNA testing in the public sphere. Media reports on these trials provided a set of narrative frameworks that became very important in framing the place and character of DNA identification technology in criminal justice, questioning the motives and value of more traditional forms of police investigation, comparing them to DNA testing with its scientific objectivity, and finding them lacking.

In what was likely the greatest media case – the 1995 trial of O.J. Simpson in California – DNA evidence played an ambivalent role. In 1994, American football star and Hollywood actor O.J. Simpson was accused of murdering his wife, Nicole Brown Simpson, and her friend, Ronald Goldman. In a televised trial lasting several months, the prosecution built its case largely around a trail of blood leading from bloody footprints at the site of the murder to the door and floor of Simpson's white Bronco, to bloodstained gloves and socks, to drops of blood in the accused's bedroom. According to the prosecution's forensics expert, DNA from a blood sample found at the crime scene matched Simpson's DNA with a 1 in 170 million chance that it was a random match. A blood sample on a sock found near Simpson's bed contained DNA that matched that of Nicole Brown Simpson with an astronomical 1 in 9.7 billion chance of a random match. However, the jury, after only four hours of deliberation, found Simpson not guilty. DNA evidence did not determine the nature or outcome of the case. In fact, the defence brilliantly managed to focus attention away from the evidence itself onto the people and processes by which it was gathered and analysed, showing their lack of care and sloppy practice, and implying that the Los Angeles Police Department was racially motivated in its DNA findings.

Although the Simpson case could be interpreted as a rejection of the supposed certainties of DNA findings in favour of other forms of evidence, such an interpretation is not entirely accurate. The case is not a rejection of DNA evidence per se, but rather a vote of non-confidence in the human element involved in the collection and analysis of that evidence. It marks a period in forensic DNA identification technology in which the procedures of collection and analysis were open to scrutiny. Other forms of expertise could intervene in its interpretation and propose alternative forms of seeing that still relied upon traditional social and legal categories of knowing. The goal and appearance of total scientific objectivity had not then been achieved. But that period is over. Courts have established the bases by which biological expertise may claim a strong authority to speak to the issue of guilt or innocence. In the United States, both public and private forensic laboratories are increasingly acquiring accreditation by following the protocols set out by the American Society of Crime Laboratory Directors. In Canada, the federal government has centralized control of forensic laboratory protocols within the jurisdiction of the RCMP. The veracity of the science is much harder to question today than it was in the early 1990s, and the collection and laboratory practices are much more tightly regulated and controlled.

As a media trial in which DNA evidence itself becomes a character in the dramatic narrative, the O.J. Simpson trial is an unsettling case, in part because its lack of narrative closure proved disturbing to much of its viewing audience. Although there was a convergence between the interests of the justice system and the media coverage, in that both expected Simpson's conviction, those expectations were foiled as the conclusion veered away from the standard script. Further, since journalistic crime narratives generally place the audience in the role of the victim of the crime, in this case, the apparent victimizer went unpunished. The outcome particularly disturbed many of the trial watchers because of their · anticipation, even in this early period of the technology, that DNA testing would simply put an end to all doubt in the case. The Simpson case was also unusual in its fact situation. Media cases in which DNA plays a prominent role more commonly centre on people who have been convicted of a crime, but who claim innocence and ultimately win the right to have their case re-examined through DNA analysis of the evidence. Such was the situation in the two most prominent cases in recent Canadian history – the cases of Guy Paul Morin and David Milgaard.

Morin and Milgaard were both accused and eventually convicted of sexual assault and murder. Both spent time in prison – in the case of Milgaard, twenty-two years, and in Morin's case, eighteen months – loudly proclaiming their innocence throughout their incarceration, and both were eventually exonerated by DNA evidence. Given the importance of these two cases in the policy discussions and media coverage of DNA technology in criminal justice, and given their prominence in the Canadian public consciousness, they have become emblematic of how DNA evidence is framed in public discourse and thus provide insight into how a particular perception of the technology is produced in the public sphere.

The David Milgaard Case

In many ways the media narratives constructed around the Morin and Milgaard cases follow a similar format; consequently, this chapter will concentrate on only one of the cases: that of David Milgaard. The events of the Milgaard case cover twenty-seven years, twenty-two of which Milgaard spent in prison. In 1970, Milgaard was tried and convicted of the sexual assault and murder of Gail Miller in Saskatoon. Much of the evidence seemed shaky, and he launched an appeal that went up to the Supreme Court, but leave to appeal was denied. His mother, Joyce Milgaard, continued to look for avenues of review of the case, but each came to a dead end until, in 1991, she approached Prime Minister Brian Mulroney on the street outside his Winnipeg hotel room and asked him to take action on the matter. He promised to investigate, and within a month, Justice Minister Kim Campbell granted a judicial review in front of the Supreme Court of Canada, a highly unusual step.[5] At this 1992 review, the main witnesses of the original trial were questioned again and found to be highly unreliable. As well, the court requested the appearance of Larry Fisher, a man who, at the time, lived in the area of the murder and who had subsequently been convicted of other similar sexual assaults. He denied any guilt. In the end, the Supreme Court went only so far as to recommend that Milgaard's conviction be overturned, the charges be withdrawn, and a retrial be held. The Saskatchewan government chose to stay the proceedings, leaving the charges standing for one year. Although Milgaard was released from prison, he was not declared innocent of the crime. This meant he was neither granted the satisfaction of being cleared nor the right to pursue compensation.

One of the means by which Milgaard attempted to demonstrate his innocence had been DNA testing. The first test on the evidence was done in 1988 when the technology first became available in Canada, but it was in its infancy and the test was inconclusive. A second test was held during the Supreme Court review in 1992, with the same result. Finally, in 1997, a third and final test was conducted on the last bit of DNA evidence from the crime scene. By this time, the technology allowed for a more refined test. That test took place in the Forensic Science Service laboratory in Yorkshire, England, conducted by scientific representatives for Milgaard, the RCMP, the Saskatchewan Justice Department, and the British laboratory. They found semen stains on the victim's underclothing, as well as on her nurse's uniform. Authorities had brought along blood samples from both Milgaard and Fisher. The Fisher sample had been taken in 1992, during the Supreme Court hearing. Fisher did not specifically consent to the 1997 test, but the federal Justice Department approved of its use. Milgaard's DNA clearly did not match the crime scene sample. Fisher's did. David Milgaard was fully exonerated as a result of DNA testing, an outcome the Milgaards had been fighting for for twenty-seven years. The Saskatchewan government admitted its mistake and began negotiations for compensation, which ended up equalling $10 million.

Constructing the Media Narrative of the David Milgaard Case

Generally speaking, press coverage of the Milgaard case has tended to avoid sensationalism of the sort associated with American media trials, particularly celebrity trials like the O.J. Simpson case. Yet there is some degree of sensationalism due to Milgaard's wild youth, in which he had engaged in unusual sex acts, taken illicit drugs, and committed petty crimes. While in prison, he had escaped twice and attempted suicide. These facts were not pertinent to his case at the beginning of his review in 1991 but are often cited in the press to add an air of personal interest. As a result, even after Milgaard's release from prison, the press has followed his personal life, reporting on a traffic violation, a speech he gaves at a high school, new relationships, and the costs of his incarceration on the lives of his brother, sisters, and parents.

Another element in the characterization of Milgaard and others involved in the case was reference to similar trials. This was not difficult in the Milgaard coverage because of the media history of the Morin case, in which some of the same people were involved. Morin himself made an appearance, having befriended Joyce Milgaard through a support group for the wrongly convicted. He was reported as being the person who informed Joyce about the results of Milgaard's 1997 DNA test results. Also, Morin's lawyer, James Lockyer, became Milgaard's lawyer in the pursuit of compensation from the Saskatchewan government and in securing a DNA test to exonerate him. There were constant references throughout the coverage to other famous Canadian trials, such as the Morin and David Marshall cases, as well as those of Rubin 'Hurricane' Carter, Steven Truscott, Thomas Sophonow, Wilson Nepoose, Susan Nelles, and Richard Norris.[6] A number of lesser-known American cases received mention as well. All of these added up to a media narrative about an apparently widespread problem that needs to be addressed.

A number of characters were important in the construction of the Milgaard narrative. David Milgaard himself was a difficult person to characterize. The press clearly labelled him as an innocent victim of an abusive system. However, he had questionable traits that rendered him ambiguous. His past suggested that he was not an innocent *person*, even if he was innocent of Gail Miller's murder. There were also reports of jailhouse confessions to the crime, although the source of these confessions was an alcoholic prison guard who eventually quit his job (Karp and Rosner 1991). After his release, Milgaard was vengeful and had a few encounters with the law. Regardless, the press could dismiss these as the adjustments of a man who was incarcerated as a youth and never properly socialized.

Joyce Milgaard was also a mixed character in the narrative. Her personal background received almost as much attention as David's, including how she had to quit school and work at a number of jobs in her youth. After David's conviction, she once again took on a number of jobs to pay for her campaign to free him, in the meantime losing her marriage, her wealth, and the stability of her family. Her crusade took on a certain fanatical characteristic, enhanced by her belief in Christian Science and a strong devotion to God.

Two other crusaders figured prominently in the case. Hersh Wolch, Milgaard's chief lawyer, took on the case pro bono and pursued it with the greatest vigour. Along with Joyce Milgaard, he explored every possible avenue to bring about a review of the case, and to proceed, after the review, to exonerate Milgaard and defend him in case of a retrial. The other was Reverend Jim McCloskey, a romantic figure, an American prison chaplain who headed an organization to reopen the cases of potentially innocent convicted persons, and who took it upon himself to advocate for Milgaard.

On the other side was the Saskatchewan government and the federal minister of justice, both of whom seemed motivated by a self-centred need to appear infallible in how their agents – prosecutors and police – conducted the investigation and trial. The minister of justice was arrogant in her refusal to even consider the matter; apparently, the twenty-three-year incarceration of an innocent man was beneath her notice. The Saskatchewan government was petty and mean in its refusal to admit a mistake and voluntarily correct it. Chief Justice Lamer, who came to represent the face of the Supreme Court in the review, went some way to recuperate the system by proving himself, within the narrative, to be interested in justice, asking penetrating and forthright questions that challenged the participants to tell the truth. However, the Supreme Court lost some credibility when it refused to exonerate Milgaard: the system continued to protect its own.

Finally, always in the background of the case, was Larry Fisher, the ominous presence, who, from the perspective of the journalists, was the obvious choice for the real murderer. Reporters provided details of his past, interviewed his ex-wife and his mother, and listed timelines of his criminal activities to imply that he was the actual culprit. Through it all, Fisher remained silent and refused to acknowledge his guilt, contributing to the impression that he was a cold-blooded psychopath without remorse for his victims, including David Milgaard.

This was the cast of characters whose interactions provided the basis for the narrative as it unfolded between 1991 and 1997. Each was a recognizable type within media narratives where abuse of power was the dominant theme. Such narratives produce a relatively unambiguous drama of crusaders versus impassive authorities, David and Goliath, the search for truth versus the misuse of authority. In the end, justice was restored, but only through the intervention of media pressure, extra-legal strategizing by the crusaders, and, of course, final arbitration provided by DNA testing. The fundamental problem of discretion in criminal justice and abuse of that discretion remains, with the implication that the next victim of this flawed and subjective system may not be so lucky.

In the end, the media narrative formed a morality play. By 1991, Milgaard was clearly innocent in the eyes of the press, and this clear, unambiguous stance made it difficult to understand the reaction Milgaard's appeals received from authorities. As the story went, Reverend McCloskey, Joyce Milgaard, and defence lawyers uncovered compelling new evidence and found solid reason to challenge the existing evidence, and yet the state would not hear it. At no time was the position of the government, police, and prosecutors presented in the coverage, and their silence gave the appearance of arrogance. The only comment reported from a government official came from William Corbett, a senior counsel to the federal justice department, who compared those who believed in Milgaard's innocence to those who believed that Elvis was still alive (Tyler 1997). In light of the apparently overwhelming evidence, however, such a comment seemed ludicrous and callous. Likewise, the decision of the Supreme Court to order a retrial and the Saskatchewan government's decision to stay the proceedings were incomprehensible within the parameters of the media narrative, wherein there was no ambiguity about Milgaard's innocence. Articles reiterated the flimsy evidence against Milgaard and the strong evidence against Fisher, whose 1994 release from prison was reported with dismay: 'A violent serial rapist accused of a murder that David Milgaard served 22 years for was to be released today from a Fraser Valley prison. Larry Earl Fisher, 44, was considered so dangerous by officials that he wasn't released a day early. He served his entire 23-year sentence for raping seven women – most at knifepoint – in Saskatchewan and Manitoba. Now there is nothing more officials can do to keep Fisher behind bars' (Hall 1994). The article proceeded to detail Fisher's crimes, implicate him in the death of Gail Miller, describe Milgaard's review, point out that Fisher denied murdering Miller, and report the shocked reaction of Joyce Milgaard to the news of Fisher's release. Not only were officials covering up their mistakes in relation to Milgaard, then, they were not pursuing the dangerous offender who committed the murder and were hampered by their own system in keeping the public safe from him.

The morality play went on to detail how the Saskatchewan government played politics in order to avoid having to compensate Milgaard and admit its mistake. It called upon other state agents – the RCMP and the attorney general of Alberta – to conduct an investigation into any wrongdoing or cover-ups in the original investigation and trial. Not surprisingly, the investigation found no evidence of wrongdoing. Joyce Milgaard was quoted as saying, 'It somehow seems wrong. The whole thing has been a whitewash and a coverup ... It's real dirty pool that they're announcing the findings of this right now when we're having a civil action before the courts ... What would you expect from Saskatchewan? They've never done right by him to begin with.' The Saskatchewan attorney general was quoted as glibly stating, 'You just have to turn the page on this and consider that the book has been completed and is now closed ... Life goes on' (Jang 1994). The Saskatchewan government appeared satisfied with how it had handled the matter and seemed to have used questionable tactics to avoid admitting its mistakes, paying compensation, and dispensing justice.

The power interests of the state and its need to appear in control of criminal justice prevented it from being able to admit the truth and restore justice. At this point in the narrative, only one force could reverse this outcome – the intervention of objective science in the form of DNA testing. After the Saskatchewan government was found not guilty of any wrongdoing, there were renewed calls in the press for DNA tests, but it took another three years for the tests finally to occur. Before then, DNA testing had been referred to on many occasions: in 1988, when Joyce Milgaard and David's lawyers initiated the reopening of the case; in 1992, during the Supreme Court hearing; in 1995, after the findings of the investigation into the Saskatchewan government's conduct in the case; and again in 1997, when, after pressure from Milgaard's lawyers, the federal Justice Department agreed to pay for tests. In each of these periods, DNA testing held out hope for the final determination of truth. In the end, it restored justice and found the truth, leading, ultimately, to exoneration and compensation for Milgaard in 1999, and to the trial and conviction of Larry Fisher, in 2000. DNA testing was necessary because it was the only avenue remaining to a clear resolution of the narrative. Milgaard's innocence could not be left unresolved. Unlike the Simpson case, where there was some doubt over forensic science in general because of the personal failings of individual forensic scientists, there was no hesitation about the utility or efficacy of DNA testing in this case. It was the final arbiter, and, by 1997, any lingering doubts about its accuracy, usefulness, and necessity, were gone. The laboratory errors of the O.J. Simpson case were forgotten. DNA testing is an objective oracle of truth in criminal justice, superior to the politicized power games of self-interested state agents.

Contrary to what many media analysts report about the normal relationship between the press and the courts, the coverage of the Milgaard case, along with that of Morin, does not represent a convergence of interests between the two institutions; instead, it manifests rupture points in that relationship. Although justice is restored within the narratives of the two cases, the problem of abuse of power and trust remains. In the media coverage, journalists put the criminal justice system on trial and find it guilty. The charge is a deficit of objectivity in the pursuit of truth and justice; authorities have too much discretion, they abuse it to condemn innocent people and subject them to torment. Reporting from this perspective places the media consumer in an unusual position. Rather than interpreting the case from the position of the victim of crime, the consumer is invited to renegotiate his or her position as one of the victims of the justice system. The claim of prosecutors and judges to be the representatives of the public within the system is painted as false. Although justice is dispensed in the end, only the intervention of the media and science brings about a positive outcome.

These conclusions flow out of a shared narrative format in the coverage of Milgaard and other similar cases. A horrific crime occurs – a young girl / woman is raped and murdered. The police investigation turns up few clues, until someone provides questionable circumstantial evidence that points to a suspect. Further inquiries turn up more equivocal evidence. Hurrying to close the case, the police arrest the shocked suspect. During the trial, he is betrayed by friends and neighbours, who are under police pressure to testify against him. Through this shaky evidence and aggressive state prosecution, the defendant is eventually convicted and sent to prison. Crusading friends, family, and lawyers mobilize to free the accused and exonerate him. The media take up the cause. To protect its reputation and the reputations of its agents, the state resists. But a DNA test provides incontrovertible evidence, and the accused is exonerated. Thanks largely to DNA evidence, which is not subject to contestation, the state must now provide justice in the form of an acquittal, an apology, compensation, and perhaps an inquiry.

Like the CSI effect, this narrative elevates science to a determinative role in justice. It is an expression of the desire for certainty of both guilt and innocence; at the same time, it provides a critique of other forms of evidence gathering, and of those who gather that evidence. Arguably, this narrative also holds an implied critique of bodily integrity. It is difficult to sustain a defence of that right from a moral perspective when a killer is on the loose, and public safety and the desire for justice are at stake. Even though Larry Fisher was not asked if he would allow testing of his blood sample in 1997, and even though no warrant was issued for a DNA sample, an old sample was used, and no one questioned this action. Fisher is a terrifying figure and ultimately guilty, but in the determination of his guilt, an unnoticed, further erosion of bodily integrity has occurred.

Genetic Justice in Canada

Representations of genetic surveillance in popular culture and press coverage, as well as its particular framing within law and policy, have played a part in the epistemological shift underway in cultural perceptions of the body and its relationship to society. Within popular culture, developments such as the CSI effect have significant governmental impact on the entry of genetic surveillance technologies into society. In the rush to further enhance the place of biotechnology within criminal justice, there seems little time to reflect upon the more cumbersome questions about the informating of genetic codes, and their addition to the information superpanopticon. The CSI effect has a depoliticizing influence on the technology, through reference to its objectifying characteristics and through its normalizing effect – making it appear ubiquitous, desirable, even unassailable. Consequently, the legal principle of bodily integrity is easily replaced by police powers to obtain DNA samples – the public and the police need to know. Objections by legal elites and privacy advocates can be dismissed as inappropriate in the quest for public safety.

Ultimately, popular culture representations of DNA identification enhance the perception of genetic surveillance and its development as a legitimate, objective, and normal part of criminal justice, despite the fact that it is a social science fiction, a simulation of surveillance. It is a social science fiction in the sense of portraying a perfected vision of genetic surveillance as a present reality, and it is a simulated form of surveillance in that it does not involve the monitoring of bodies, but rather of abstracted information from bodies, which can be manipulated and transmitted at the speed of light. Forensic scientists who express a desire for the technologies, resources, and access portrayed in fictional representations are asking to close the gap to allow them to achieve the imagined end point of the technology. In this way, the representations act as a biogovernmental normalization of the simulation of genetic surveillance, and a shift in the citizen's sovereignty over his or her body.

Within the press coverage of the David Milgaard case and other media trials, two recommendations are offered to resolve the problem of subjectivity in criminal justice. The first is additional layers of bureaucracy, which would audit the decisions and actions of other layers. The second, and argued to be much more effective, is DNA testing. Press reports reify DNA as something that transcends human agency to form a force for justice in its own right. There are few doubts expressed about its efficacy, and the narrative appears to pin all hopes on this technology to rebalance the justice system in dealing with violent crime. The argument is compelling and, like the CSI effect, has its basis in the biotechnological imaginary. Proponents of DNA testing have a number of good examples of its benefits, not only in freeing the innocent, but also in convicting the guilty, such as Larry Fisher. Those who are hesitant about its implementation due to possible erosions of privacy, increases in social surveillance, and potentials for genetic discrimination have few cases of technological abuse to support their position. DNA was determinative in the Morin and Milgaard cases. The charisma of genetic science led to an immediate acceptance of the DNA verdict in both instances, and to government apologies and compensation. It restored justice. In this way, DNA testing technology becomes an antidote to politics, rather than an object of politics. It is depoliticized through the framing of the media trial narrative. Although science and scientists are often the targets of calls for public scrutiny and democratic accountability in late modern risk society, technology seems to remain immune to politicization. In Canadian criminal justice, politicization occurs around the risks implied by rising crime rates and abuse of expert discretion. DNA technology is a means for bringing both criminals and crime experts under control by objectifying crime control. This is its power within the biotechnological imaginary; it adds a utopian element to the media narrative, without discussion of what is involved in placing genetic surveillance in such a prominent place in criminal justice.

How has the biotechnological imaginary worked itself out in the sphere of public policy? This is also an important public site for evaluating the nature of genetic surveillance in society – press and popular culture form part of the enabling conditions for practical and ideological developments in genetic surveillance. In terms of policy developments, there are three interesting indicators of how the depoliticization, rationalization, scientization, and reification of DNA and genetic surveillance, which we see in popular culture, are operating in governmental and police practices. First is the policy consultation that preceded the DNA warrant databanking legislation. During these consultations, when stakeholders participated in setting the boundaries of debate over the DNA warrant and banking provisions, the primary narrative frame, which developed in locating this new technology within existing meaning systems, was 'privacy.' The main reasons for this framing lay in the ways in which the language of privacy was used in consultation questions set by the Department of Justice, and in the cautionary voice of the privacy commissioner, whose responses were located in terms of privacy: 'Retaining a databank of genetic samples from convicted offenders will inevitably attract researchers who want to analyze the samples for purposes that have nothing to do with forensic identification. This scientific curiosity, coupled with growing pressure to reduce crime by whatever means, no matter how intrusive, will almost certainly lead to calls to use samples to look for genetic traits common to "criminals." This type of research, while perhaps of scientific interest and possible social value, raises complex legal, ethical and moral problems that we have yet to resolve' (Privacy Commissioner 1998, 4).

A number of fears are expressed here: that concern over crime rates leads to a reduced concern over individual privacy, anxieties over the absence of moral constraints on scientific research, and, finally, the recognition that continuing genetic research increasingly informates the human body and the human person – illuminating and codifying the fundamental building blocks behind physical and personality attributes. In response, the Canadian Police Association, which advocated a much wider use of the technology, stated, 'The argument against using the broad base gathering [of DNA samples] ... is based on "what ifs" and potential abuses imagined by those who oppose the use of this tool. No doubt, had the Office of the Privacy Commissioner existed 100 years ago, similar "objections" would have been trumpeted to fingerprinting' (Canadian Police Association 1998, 3). In the policy process, however, the issue of privacy was eventually resolved by converting it from an ethical question to a technical one – privacy safeguards can be included in the legislation to satisfy the *Charter of Rights and Freedoms*. The *Charter* became a mechanism for resolving the debate.

The second important policy decision in the development of genetic surveillance in Canada lies in the language of the legislation that enacts the DNA warrant provisions. As noted earlier, Bill C-104, passed in 1995, empowers the police to use as much force as necessary in seizing DNA samples from suspects for whom a warrant has been granted. What is interesting about this provision is the absence of a subsequent clause requiring the police to do so while respecting the bodily integrity of the suspect. The only other provisions that allow police to seize bodily substances are in sections 254 and 256 of the *Criminal Code*. Section 254 empowers police to demand a breath sample from anyone operating a vehicle who is suspected of having alcohol in his or her body. Under this section, police can also demand a blood sample, but only under very strict conditions.[7] Section 256 also allows blood sampling in impaired driving cases and is meant to cover situations in which there has been an accident and the suspect is unable to give a breath sample. In both sections, the offence must have occurred within the previous two hours; blood samples can be taken only if breath samples cannot be obtained; and a medical practitioner must be present. Most importantly, officers are cautioned that they must respect the bodily integrity of the suspect. Unlike these other sampling provisions, the DNA warrant legislation uses a language of privacy protection for the genetic information taken from a suspect, but there is no deference to the principle of bodily integrity. As a result, legal experts at the time predicted it would be struck down in court. Yet, in a series of landmark cases, it was upheld.

The third significant element of the DNA warrant and banking provisions involves an increasingly significant practice in criminal investigation – the police tactic of DNA dragnets. These are police investigations in which people are asked to submit to mass DNA testing to exclude themselves from the suspect list. There have been four major DNA dragnets in Canada: Vermilion, Alberta (1994), Port Alberni, BC (1996), Sudbury, Ontario (1998), and Toronto, Ontario (2003). In the Vermilion case, which is fairly typical, men in the town were asked to submit to DNA testing in an effort to identify a rapist who had been operating in the area for three years. No one refused because of the obvious stigma attached. An RCMP spokesperson commented, "'I'm sure if someone were not to give blood and that were found out, he would be really, really unpopular'" (Plischke 1995). A number of men in the town reinforced this view by telling reporters that they would not look favourably upon someone who refused to give a sample. After eighteen months with no results, the police called a town meeting to motivate the town to continue to give samples. The mayor stated, "'Anyone who wants to protect the system and wants to live in a safe society, why would we have anything to hide?'" (Plischke 1996a). At the meeting, two men spoke out against the mass testing as a violation of privacy, but were shouted down by other residents who supported the investigation. Afterward, the police praised the community for their cooperation and warned that anyone who did not give a blood sample on request would face an intrusive background check (Plischke 1996b). Despite these tactics, the case has never been solved.

Defining the primary political and ethical issue as privacy, absenting any reference to bodily integrity in the legislation, and conducting DNA dragnets engenders material traces of a larger biogovernmental shift in definitions of the body, and how the body relates to state authority. This entails more than the collection of information about persons. It involves informating the body in ways that may change identity and embodiment. As discussed earlier, translation of the body into digital code amounts to a change in the level of ontology rather than merely representation (van der Ploeg 2002). The familiar anatomical-physiological body of flesh and organs is itself a contingent historical construction, which results from the late-eighteenth-century combination of anatomy and experimental science, and is not always the natural pre-discursive referent that it is supposed to be. The common law right of bodily integrity is attached to this understanding of the body. It is the physiological body that is protected from the power of state agents in the criminal context, through a duty of those agents to observe bodily integrity in exercising their special powers – a duty not to enter the body and to avoid violating the sovereignty of the individual over his or her body, to whatever extent possible.

Nevertheless, during the course of the policy and implementation process of the DNA scheme in Canadian criminal justice, bodily integrity was elided from the discussion and replaced with a language of privacy. This demonstrates shifting notions of the body from anatomical object to information. Privacy relates to protecting information. Bodily integrity relates to protecting the thing itself. If the body is defined as genetic information, it is opened up to protection based on information privacy. This is what occurred in the policy process around the genetic justice system, and it is what is occurring in popular culture representations. It is a process of normalization for the redefined, informated body – a change that is enshrined in legislation that excluded bodily integrity but included privacy safeguards. Consequently, the locus of the right has shifted. It no longer occurs at the point of entry into the citizen's body, but afterwards, when the state comes into possession of genetic information. It is that information that is protected. This reduces the sovereignty of the citizen over his or her body and empowers the state. The conditions are now present to open the subject up to the molecular gaze of genetic surveillance with the full cooperation of the citizenry.

The DNA dragnet demonstrates an outcome of the redefinition of the body in informational terms. If the objection is one of privacy and the police can guarantee genetic privacy, there is no more moral ground upon which to object to state DNA testing on demand. If you have done nothing wrong, you have nothing to worry about. In other words, viewing the issue as something other than the freedom to manage one's own body means that there is an accompanying shift in governmental language from rights to responsibilities. This is how biogovernmental responsibilization enters into the genetic surveillance system. Privacy is not so much a right as it is a responsibility of the state towards the citizen. In return, the good citizen has a normative responsibility to submit to genetic surveillance in order to make criminal investigation more efficient, and to do his or her part in protecting society. What occurs, then, is a distribution of responsibilities, rather than an assertion of rights. The mayor and citizens of Vermilion, as well as the RCMP, made this point by warning residents of Vermilion that they had a responsibility to submit to DNA testing and had no legitimate excuse to refuse. It also appears to be the point behind the CSI effect and the press's enthusiasm for more DNA testing, following the outcome of the Milgaard case.

Accompanied by media trial narratives and television dramas, the actual implementation and administration of genetic surveillance has entered the public sphere in a relatively smooth and unproblematic way. Despite initial fears by policy stakeholders that there might be a public reaction against such a powerful technology of genetic surveillance, public acceptance suggests that citizens have embraced it. That genetic surveillance has been directed only toward criminal offenders makes it seem benign and even a social good. The ease with which it is normalized in Canadian society is an indicator of public fears of crime and terrorism, people's comfort with surveillance, and the charisma of genetic science. The hopes and fears surrounding these scientific promises can be put to use by authorities, who are currently exploring the limits of biogovernance. The genetic turn in current dominant forms of body ontology has the effect of redefining the body as information, which then redefines entry into the body as a mode of information retrieval, justified by a governmental language of individual responsibility for controlling the risks of modern life. It is probably impossible to return to a notion of bodily integrity rights within this discourse; genetic science has changed notions of public and private, internal and external, in relation to the body. As a result, some notion of integrity must be developed that is appropriate to contingent bodies, which are open to biogovernance.

Biogovernance, Genetic Surveillance, and the Biotechnological Imaginary

The idea of genetic surveillance seems to hold a certain fascination in twenty-first-century North American societies, and that fascination will only continue to grow. Forensic scientists are calling for more personnel, more technology to catch up to their fictional colleagues, and fewer institutional constraints on their work. The media are calling for an expanded genetic justice system as a remedy for abuse of authority within the criminal justice system. Police are calling for a greater degree of surveillance by loosening constitutional restrictions on who can be entered into the National DNA Data Bank in Canada. All of these calls are ultimately about closing the gap between the body and the code and moving toward an imagined perfect system of surveillance that will bring about a perfect security. At the same time, genetic surveillance has captured the imagination of the public, which appears to be signalling its readiness to submerge itself into the superpanopticon, while also becoming its functionaries by flocking to forensics training programs.

All of the above suggests that there is great potential for expansion of genetic surveillance as a biogovernmental process; the logic of the biotechnological imaginary makes its spread likely. Political questions about this growth as an extension of state power over the citizen's body are lost in the public's fear of crime, faith in genetics, and comfort with surveillance. Objectification of the gene as a form of information and as an object of simulated surveillance feeds into the social science fiction of instant observability, as represented in popular culture narratives. These same narratives contribute to a normalization of genetic surveillance, along with journalistic accounts of trials and public policy processes. Each of these sites of meaning-making draws upon and feeds into the biotechnological imaginary with the effect of legitimating the new technology and rendering it essential for public safety. However, public safety can be guaranteed only if every citizen is responsibilized into participating in the genetic surveillance system. By redefining the body as a site of identification, whose privacy can be guaranteed by the state, there is no legitimate excuse to refuse government agents access to the body – a small step from requiring all citizens to put their DNA on record. In this way, genetic surveillance, although it has entered into society in a relatively quiet manner, has been an extremely important development in biogovernance. In a few short years, it has changed the balance of power between the citizen and the state, and that shift will likely proceed further as police forces, forensic scientists, the media, and perhaps the general public lobby to allow a loosening of constitutional restrictions on the operation of genetic surveillance. Canadian society, at the turn of the millennium, is only one remove away from a full genetic justice system – one in which the body is not simply a site of information, but also of prediction.

CHAPTER THREE

The Sexual Politics
of Biotechnology

'I used to think of my body as an instrument, of pleasure, or a means of transportation, or an implement for the accomplishment of my will.'

Atwood (1998, 91)

Human beings hope they can stick their souls into someone else, some new version of themselves, and live on forever.

As a species we're doomed by hope then?

Atwood (2003, 146)

Margaret Atwood offers us two radically different, and yet equally unsettling, visions of our sexual future. In *The Handmaid's Tale* (1998), sexuality, reproduction, and technology intertwine to tell the story of Offred, a young woman forced into reproductive servitude under a fascist state in the Republic of Giliad. Rights of motherhood are given only to women who are aligned with the scientific and military elite, regardless of their physical ability to actually bear children. This gap between social and biological roles is filled by the handmaids: fertile women without the social status to be mothers. In *Oryx and Crake* (2003) women's wombs are obsolete, replaced by laboratory-bred hybrid humans produced by messianic, mad scientist, Crake, in a world dominated by transnational global capital and spin. His Paradice Project, produced by underground genetic scientists, produces a new species of human engaging in sexual reproduction without emotion; thus the new hybrid humans are perfected 'hormone robots,' who have sex only to procreate, the ultimate endpoint of sociobiology.

While both novels are speculative fiction, the handmaids and their society serve as a cautionary tale, a political parable of a misogynist future, whereas Crake's vision manifested is an extension of our current biotechnological enterprise, a critique of capitalism and science out of control. The first novel is told from the perspective of a woman, the second from that of a man. The first focuses on women's bodies as social lodestone; the second effaces women's reproductive capacity and relocates it in the hands of the scientist. The latter novel, therefore, is hi-tech, the former, quaintly analogue in its treatment of reproductive technology. In *The Handmaid's Tale* the state is a central social actor, seeking control over reproduction and hence women's bodies. By contrast, in *Oryx and Crake* the state has been expunged as a relevant agent, replaced by markets and capital, and the charismatic technoscientists in their service.

Atwood's novels effectively highlight changes in how post-industrial societies understand human reproduction, the proper place of technology in its development, and the role of women, scientists, economy, and governments in its governance. We suggest that in the Canadian state's encounter with the sexual politics of biotechnology similar shifts are visible. As reproduction is brought within the realm of the governable, there is a shift away from the primacy of women's bodies, an increased role for scientific enterprise and industry, a declining role for the state, and the entanglement of reproduction in the prognostications of biotechnology. Correspondingly, we also see shifts in the discourses of governmental legitimacy, from women's bodily autonomy to the support of fertility as an abstraction. This chapter explores the official discourses of biotechnology in Canada as the state lays claim to managing reproduction as a social, as well as biological, function.

To do so, we examine four key moments in Canadian legal and political history. Each focuses our national gaze on reproductive technologies, which we define here as any technology that manipulates animal reproductive matter, such as embryos or fetal tissue. Each event is a rupture point, an explosion of social forces highlighting competing visions of women's sexuality, reproduction, and biotechnology. Change is demanded and the state is expected to act. Yet how, whether, and when the state responds and within which rhetorical frameworks of responsibility are always undetermined and unpredictable.

The first moment is the decision of the Supreme Court of Canada in 1988 to strike down Canada's abortion laws. As an early case under the newly minted *Canadian Charter of Rights and Freedoms*, this case was more than a victory for pro-choice advocates; it also dramatically transformed the legal and political landscape in Canada. It set the stage for the second key moment, one year later, a battle that once again pitted abortion supporters and foes and brought another case to Canada's highest court. This time the Supreme Court was forced to rule on the rights and ownership of the fetus in contests between women and potential fathers. While these two sets of legal disputes are not strictly about genetic reproduction, the issues of women's reproductive rights and the status of embryos vis-à-vis the women who carry them are at the heart of later policy initiatives on genetic reproductive technologies. They set key discursive frames within which subsequent policy-making takes place, with which it must engage, and which it ultimately rejects.

In the face of its (deliberate) failure to draft legislation on abortion, the Canadian government turned its attention away from the issue of women's reproductive rights and in 1989 established the Royal Commission on New Reproductive Technologies – our third event. After four highly acrimonious and controversial years, the commission produced a massive, thirteen-volume report entitled *Proceed with Care: Final Report of the Royal Commission on Reproductive Technologies* (1993). It was, however, not until eleven years later, and after four failed attempts, that the federal government finally passed legislation in 2004 governing the use of embryonic and fetal matter in biotechnological research and commercial enterprise. This legislative saga constitutes our fourth and final moment in the governance of biotechnological reproduction.

Atwood suggests her novels are works of 'speculative fiction,' that they explore the realm of the imagination, the relationship of humanity to the universe, the interlinking of nature and technology, and the reshaping of society to contend with these dramatic transformations (2004, 515). Each of our four legal-political events does the same symbolic work as Atwood's speculative fictions, culminating in legislation that is, itself, a social science fiction, regulating a range of current and future reproductive technologies and signalling a shift in governance from bodily discipline to biopolitics.

A Context of Competing Interests in Reproduction

The Canadian legal debate over contraception and abortion began in earnest in the 1960s when the intersection of a number of related but separate events led to a full-scale revision of contraceptive values that demanded new forms of governance. The birth control pill, which had been in development in the United States since 1950, was commercially released as Enovid in 1960 by Searle Pharmaceuticals (see Asbell 1995; McLaren 1999; Tone 2001; Watkins 1998). In Canada at the time, it was still illegal to sell or advertise contraceptives under section 207 of the *Criminal Code*, with penalties of up to two years' imprisonment. However, the law also provided that convictions would not be pursued if 'the public good was served' (Watters 1976, 121–2). The 'public good' was at the heart of any rationalizations for the development and marketing of oral contraceptives. The pill's development had been funded and supported predominantly by members of family planning associations, which were interested primarily in population control and the limiting of children for the poor or 'unfit.' Through the discourse of responsible parenthood, birth control was hailed as a way to engineer an improved society. It promised the control of certain 'less desirable populations,' and could foster, as the Canadian Family Planning Federation stated, 'good citizenship through responsible family life' (Appleby 1999, 45). Having children became increasingly a question of planning, programming, and research, a set of values that eventually and inevitably came to mark conception as much as, if not more than, contraception.

However, the 'contraceptive revolution' of the early 1960s intersected a very different sexual revolution. The former was focused upon families, nations, and scientific progress, whereas the latter celebrated the individual and her consumer-based erotic pleasure (Watkins 1998, 2). By the late sixties, the women's liberation movement was contesting both these earlier versions of women's sexual emancipation. They raised suspicions about claims of women's newfound sexual freedom and the efficacy – not to mention safety – of those contraceptives that were supposedly at the root of that freedom (Watkins 1998, 5). Angus McLaren contends that 'the pill' offered a medical entry into the world of fertility management, in that it chemically altered the body's cycles. As he states, 'The appearance of the contraceptive pill – not a conversion to feminist, neo-Malthusian or eugenic arguments – brought the medical profession onside in favour of birth control' (1999, 168). Given that the main sources of research funding were heavily invested in social policy initiatives to promote a tightly controlled notion of the nuclear family, as well as to secure the family's status as the primary subject position of the nation, the situation seems more complicated than McLaren implies. However, he is correct in recognizing the increasing role that the medical profession wished to play, and was playing, in the control of conception.

The oral contraceptive and the intrauterine device, a metallic device that required careful insertion directly into the uterus by a trained professional, were the two systems of contraception control preferred by the medical profession. As both methods altered women's biochemical balance, doctors were able to define contraception as a medical procedure, thus necessitating careful surveillance to ensure compliance, under the guise of ensuring efficacy and safety for women.[1] The Canadian Medical Association (CMA) insisted that, in order to guarantee proper usage of such techniques, the medical community should retain authority over contraception and be provided with a clear set of regulations – not to protect the women using the technologies, but to protect medical practitioners from malpractice suits over birth defects, miscarriages, or maternal deaths (Appleby 1999, 30). Thus, the state was being invited to provide a legislative support framework for the medical profession's control of conception.

In 1966, the Liberal minority government under Prime Minister Lester B. Pearson convened a hearing of the House of Commons Standing Committee on Health and Welfare in order to discuss four private members' bills on birth control (Appleby 1999, 16–17). Representatives from the CMA were among the first to present to the committee, noting that four years earlier, in contravention of the criminal law, they had begun accepting advertisements for, and articles about, contraceptives in their journal (30). Despite public opinion that, regardless of its disputed moral status, the distribution of contraceptive devices should not be a criminal act, it took three more years for Parliament to actually modify the laws.

In 1969, Parliament under the leadership of Justice Minister Pierre Trudeau decriminalized birth control with the passage of an omnibus bill, Bill S-15, designed to bring into effect a variety of changes to criminal legislation. Birth control would now be subject to regulation under the *Food and Drugs Act*, rather than the *Criminal Code*. However, the legislation continued to regulate the advertising of contraceptive devices and products in order to ensure there would be no 'offence to public morals' – in other words, to prevent their marketing to single women, minors, or those who might use them exclusively for sexual pleasure. Because no provinces had implemented birth control clinics or family planning services, access to birth control was primarily through family physicians and thus the prerogative of middle- and upper-class married couples in urban centres.

Interestingly, on the same day that it decriminalized birth control, Parliament also altered the legislation criminalizing abortion, bringing it, too, within the purview of the medical establishment. Bill C-150 amended section 251 of the *Criminal Code*, clearly stating the conditions under which an abortion could be procured legally. Abortions would now be legal, but only at accredited hospitals, wherein a medical tribunal of at least three doctors had been convened and had determined that the abortion was necessary to preserve the life or health of the pregnant woman. Women requesting abortions were required to plead their case before this tribunal, which held the final decision-making authority.

With legislative action establishing medical authority over both contraception and abortion, reproduction as a problem of governance was seemingly resolved. Agencies such as the Family Planning Federation of Canada were left to ensure access to birth control and abortion across the country. However, since no state resources were allocated to set up a nationwide system, access to birth control and abortion varied widely among the provinces for ideological and material reasons. Abortion and birth control therefore remained the privilege of the urban middle-class; rural and poor women struggled to receive proper information and care. Further, because hospitals were not required to establish a tribunal, some provinces simply refused to do so, in effect rendering moot abortion's decriminalization (Appleby 1999, 219). The legislation clearly determined the object to be regulated as women's bodies and their capacity to reproduce. However, authority for the control of reproductive capacity was delegated to the medical profession and effectively re-moralized through the exercise of provincial jurisdiction over health and local hospital boards' discretion.

R. v. Morgentaler: Women's Bodily Sovereignty

Downloading the responsibility for governing abortion onto the provinces and the medical establishment had created a situation where many women pursued illegal abortions in less than safe conditions, and many were denied access altogether. However, some Canadians began to crusade to overturn the legislation on human rights grounds. Dr Henry Morgentaler was among them, having first become involved in the issue when he presented a brief to the Federal Standing Committee on Health and Welfare, when it was considering decriminalizing abortion in 1967. In 1969, he brought his general medical practice to an end as his Montreal clinic started to specialize in abortions.[2] Morgentaler's clinics were open to any woman who requested an abortion, and while they charged a nominal fee to cover expenses, they otherwise made no claims on their clients. Although there were other doctors who also offered abortions, Morgentaler was the most public and daring, rocketing both himself and the issue of a woman's right to an abortion (and a doctor's right to perform it) into the public eye.

His Montreal clinic was raided by police in 1970 and he was criminally charged in both 1970 and 1971. In May 1973, he finally decided to force the issue. After boasting in the press that he had performed at least five thousand abortions already in his career, Morgentaler invited television cameras into his Montreal clinic to broadcast an abortion being performed, clearly flouting the authority of the state. This led to an inevitable third raid and set of criminal charges. His first case went to trial, but a jury of his peers acquitted him on 13 November 1973. A frustrated Crown appealed the acquittal and, on 25 April 1974, the Quebec Court of Appeal took the rare step of overturning a jury decision, substituting a conviction. In July, Morgentaler was sentenced to eighteen months in jail but remained at large until his appeal to the Supreme Court of Canada was heard. On 26 March 1975, Morgentaler's pending appeal was dismissed and the Supreme Court held that he must serve out his sentence, making him the first person in Canadian history acquitted by a jury to be sentenced to jail time (*Morgentaler* 1975). He eventually served ten months of his sentence.

In its procedurally based judgment, the Supreme Court of Canada flatly refused to take a stand on the substantive issue of the legality of abortions and how they should be regulated in Canada, denying that it had the authority to do so. 'Parliament may determine what is not criminal as well as what is, and may hence introduce exemptions in its criminal legislation. Parliament has made a judgment which does not admit of any interference by the Courts' (*Morgentaler* 1975, para. 10). The court ruling states further, 'The *Canadian Bill of Rights* must not be regarded as charging the courts with supervising the administrative efficiency of legislation or with evaluating the regional or national organization of its administration' (para. 26).

But the saga did not end there. In May 1975, while serving his sentence, Morgentaler went to trial on a second count of procuring an abortion. A jury again acquitted him and the Crown again appealed. This time, however, because the Supreme Court had rejected the Court of Appeal's right to overturn a jury verdict in its earlier decision, the Quebec Court of Appeal was forced to dismiss the appeal in early 1976. The minister of justice intervened at this point, and the 1974 conviction was set aside, leading to a new trial, where, in March 1976, Morgentaler was acquitted for a third time by a Quebec jury. With the administration of justice and the government both being brought into disrepute repeatedly by Morgentaler's flagrant challenges, and the repeated jury acquittals serving as a clear gauge of public opinion, the federal Ministry of Justice created the Committee on the Operation of the Abortion in Law in September 1975, chaired by Robin Badgley, to examine the application of therapeutic abortion law across the country. This committee confirmed in its damning 1977 report what many women across Canada already knew, that abortions were not equally available to all women in Canada. The report stated,

> The procedures set out for the operation of the Abortion Law are not working equitably across Canada. In almost every aspect dealing with induced abortion which was reviewed by the committee, there was considerable confusion, unclear standards or social inequality involved with this procedure. In addition to terms of the law, a variety of provincial regulations govern the establishment of hospital therapeutic abortion committees and there is a diverse interpretation of the indications for this procedure by hospital boards and the medical profession. These factors have led to: sharp disparities in the distribution and accessibility of therapeutic abortion services; a continuous exodus of Canadian women to the United States to obtain this operation; and delays in women obtaining induced abortions in Canada. (Committee on the Operation of the Abortion in Law, 17)

Without doubt, the federal and provincial governments were being incited to act. In November 1976, the newly elected Parti Québécois announced that all outstanding charges against Morgentaler would be dropped and further, that doctors in Quebec would no longer be prosecuted for providing abortions.

The issue might have rested there, except that in 1982 Canada's legal and social landscape transformed dramatically with the repatriation of the constitution and the enactment of the *Canadian Charter of Rights and Freedoms*. The legal branch of the state was now empowered in a way it had not been previously to oversee the actions of the legislative branch. The opportunity presented by the *Charter* was not lost on Dr Morgentaler and his counsel. Morgentaler opened a clinic in Winnipeg in 1983 and it was raided in June; however, the charges did not lead to a trial. He then opened a clinic in Toronto, which was duly raided and charges were laid against three doctors, including Morgentaler. The situation escalated politically in Toronto when arsonists attacked the clinic at the end of July.

After a variety of motions were heard by the courts, at their trial the doctors were all acquitted by a jury on 8 November 1984. The clinic reopened, but by December Morgentaler was charged again. In 1985, the Winnipeg clinic reopened in March and criminal charges were brought. Finally, the Ontario Court of Appeal heard the appeal of the Crown from the 1984 acquittal pertaining to the Toronto clinic, the Crown's appeal was allowed, and a new trial was ordered. Morgentaler appealed that decision to the Supreme Court of Canada. The appeal was heard in October 1986, and finally on 28 January 1988 the Supreme Court of Canada, in a 5–2 decision, declared Canada's abortion law unconstitutional.

Morgentaler's lawyers argued that section 251 of the *Criminal Code* was ultra vires Parliament, based on section 7 of the *Charter*, which guarantees to everyone in Canada life, liberty, and the security of the person. By forcing women to submit their bodies to the will of a medical tribunal, lawyers argued, their fundamental rights to bodily integrity were being denied. This case then was more than a question of reproductive rights, but emerged as a litmus test for the new constitution and the authority of the courts to uphold citizens' *Charter* rights against the government.

In a dramatic shift from its earlier ruling on the court's authority in relation to Parliament, Chief Justice Brian Dickson stated, 'Although no doubt it is still fair to say that courts are not the appropriate forum for articulating complex and controversial programs of public policy, Canadian courts are now charged with the crucial obligation of ensuring that the legislative initiatives pursued by our Parliament and legislatures conform to the democratic values expressed in the *Canadian Charter of Rights and Freedoms*' (*Morgentaler* 1988, 46). Pundits across the country recognized that this heralded a new era in the judicial and legislative power balance in Canada (e.g., MacQueen 1988).

On the issue of whether or not the provision violated a woman's right to security of the person under section 7 of the *Charter*, the chief justice was unequivocal:

At the most basic, physical and emotional level, every pregnant woman is told by the section that she cannot submit to a generally safe medical procedure that might be of clear benefit to her unless she meets criteria entirely unrelated to her own priorities and aspirations. Not only does the removal of decision-making power threaten women in a physical sense; the indecision of knowing whether an abortion will be granted inflicts emotional stress. Section 251 clearly interferes with a woman's bodily integrity in both a physical and emotional sense. Forcing a woman, by threat of criminal sanction, to carry a foetus to term unless she meets certain criteria unrelated to her own priorities and aspirations, is a profound interference with a woman's body and thus a violation of security of the person. Section 251, therefore, is required by the *Charter* to comport with the principles of fundamental justice. (*Morgentaler* 1988, 56–7)

The court was also very concerned with the delays in gaining access to abortion and their potential damage, physical and psychological, to the pregnant woman. It had very harsh words for the actual operation of the abortion administration system in Canada.

Given that this case decided a deeply divisive social issue that put the bodily integrity of women on trial, the recent appointment of the first woman justice to the Supreme Court was a relevant factor in the reception of the decision. Justice Bertha Wilson was clearly being evaluated as a new judge, but more specifically as a new woman judge. Indeed, her judgment offered a distinctly feminist perspective on the issue. She boldly opened with the claim that the real issue is whether women have a right to access an abortion under section 7 of the *Charter*, concluding that they do:

> I would conclude, therefore, that the right to liberty contained in s. 7 guarantees to every individual a degree of personal autonomy over important decisions intimately affecting their private lives. The question then becomes whether the decision of a woman to terminate her pregnancy falls within this class of protected decisions. I have no doubt that it does. This decision is one that will have profound psychological, economic and social consequences for the pregnant woman. The circumstances giving rise to it can be complex and varied and there may be, and usually are, powerful considerations militating in opposite directions. It is a decision that deeply reflects the way the woman thinks about herself and her relationship to others and to society at large. It is not just a medical decision; it is a profound social and ethical one as well. Her response to it will be the response of the whole person. (*Morgentaler* 1988, 171–2)

Madame Justice Wilson's language empowered women in a Kantian claim to ethical subjectivity: 'In essence, what [the existing law] does is assert that the woman's capacity to reproduce is not to be subject to her own control. It is to be subject to the control of the state ... She is truly being treated as a means – a means to an end which she does not desire but over which she has no control. She is the passive recipient of a decision made by others as to whether her body is to be used to nurture a new life. Can there be anything that comports less with human dignity and self-respect? How can a woman in this position have any sense of security with respect to her person?' (*Morgentaler* 1988, 173). She further stated that the issue was really one of 'whose conscience' – the woman's or the state's – should prevail over a matter that was, in her mind, 'purely personal and private' (*Morgentaler* 1988, 181). Therefore, Justice Wilson endorsed a strong notion of women's reproductive autonomy grounded in bodily sovereignty; the woman's body offers a boundary that the state should be very wary about crossing.

But discourses of women's autonomy were not the only ones in circulation. In his dissenting judgment, Justice William McIntyre (also writing on behalf of Justice La Forest) found anathema the idea that a women's reproductive body could be free of the state, and not part of the broader social good, as determined by the legal and medical establishments. He favoured Parliament's position of balancing the rights of the pregnant woman and of the 'unborn child,' suggesting that a woman's desire alone is not adequate reason, as Parliament had decreed, to grant a 'socially undesirable conduct' (*Morgentaler* 1988, 135–6). 'It must, surely, be evident that many forms of government action deemed to be reasonable, and even necessary in our society, will cause stress and anxiety to many, while at the same time being acceptable exercises of government power in pursuit of desirable goals ... Governments must have the power to expropriate land, to zone land, to regulate its use and the rights and conditions of its occupation. The exercise of these powers is frequently the cause of serious stress and anxiety' (*Morgentaler* 1988, para. 231). By likening the pregnant body to real estate, Justice McIntyre reiterated a major quandary in Canada's reproductive terrain: is a woman's autonomy reduced by her state of being pregnant? Thus, the court was facing a significant dilemma: can the law account for, or does it produce, a hybrid? Is the pregnant woman a less than full liberal, individual subject, or is it the pregnant body that is a split subject, with neither mother nor fetus fully autonomous from each other or the state?

The majority of the court unequivocally found only one legal subject: the woman. And while this appears on its face to then be a victory for women as legal and political subjects, closer critical analysis suggests that it may be more of a victory for the (unsexed) individual subject of liberalism. P. Lealle Ruhl (2002) notes that the law has had a long history of discomfort with the pregnant woman as a subject. Echoing Ngaire Naffine (1998), she argues that this is because 'liberal subjectivity demands an unequivocal and clear division between persons. That is to say, the liberal individual is posited on an abstraction; such an individual has no specific (meaningful) body, no history, no attachment to specificity' (41). In her embodiment and with a simultaneous, and non-competitive, imperative towards self and other care, the pregnant woman is too specific. She is simultaneously profoundly private and public. Yet the legal and political systems are ill-equipped to deal with this. Ruhl recognizes, 'In order to achieve recognition as a liberal citizen, one must first prove one's ability to separate oneself from the particularities that bodily experience invites' (2002, 41–2). And so, while Canadian women did have their bodily autonomy ratified by the courts, it was, necessarily, at the same time an abstracted rather than specifically gendered or sexed body that was protected.

The decision was rhetorically framed as also a victory for women's reproductive freedom. However, arguably the legal and medical systems combined to render women beneficiaries, rather than active agents, in a battle over the control over their own bodies. Few commentators recognized the irony. The main message in the press coverage the day following the decision was that 'Canadian women now have unrestricted access to legal abortions if they can find a doctor willing to perform them' (Vienneau 1988b) and that Canada had entered an era, until further legislation, of 'abortion on request' (Bindman 1988). Morgentaler supporters were overjoyed with the decision, but the Roman Catholic archbishop of Toronto described the decision as a 'disaster' and 'uncivilized' (Carter in Smith 1988). Morgentaler himself stated that he believed 'it's the greatest achievement that I have personally contributed to the quality of Canadian society' (in Bindman 1988).

The Canadian Medical Association, refused the authority of the court, sending telegrams to every one of its 46,000 members advising them that they should 'keep on as if the law had not been struck down' (Blaire in Vienneau 1988a). Canada was thus in a legal lacuna – abortion was not legal; it was, more accurately, not illegal. Repeatedly, social actors in the abortion battle, as well as media pundits, called for Parliament to act to resolve the issue. The federal government did hastily announce that it would not invoke the 'notwithstanding clause' to override the *Charter*, with Justice Minister Ray Hnatyshyn leaving open the question of whether the government would propose new legislation. The key question facing the government at this juncture was whether reproductive laws should be based on women's rights and bodily integrity or on women's responsibility to place their bodies in the service of the state, and act as responsible citizens under proper medical guidance. As an editorial in the *Globe and Mail* noted the day after the decision was announced, 'If Ottawa believes that society has an interest in the health of the fetus after a certain stage of development – as the court, including Judge Wilson, acknowledged it might – it can draft a law which conforms more closely with the *Charter.* Or it may choose to leave the issue alone, trusting in the medical profession and the pregnant women themselves to do the right thing' (*Globe and Mail* 1988).

While generally viewed as a victory for women's reproductive rights, the larger event of the 1988 Morgentaler case reveals the tenuous status of a biogovernance anchored in a rationale of women's reproductive agency. The court was clearly divided on how far it wanted to go; the legal decision clearly did not resolve the moral issue; citizens were taken aback by the implications of the *Charter* for the balance of power between Parliament and the judiciary; and an expressly feminist rationale for the decision appealed to a relatively narrow profile of Canadians. The issue loomed, inevitably, over the upcoming federal election.

Further, while the Supreme Court had been able to contain the subjectivity of the fetus in endorsing women's rights (it was not necessary to the disposition of the appeal to rule on whether or not the fetus had a section 7 *Charter* right), it was not going to be able to avoid that issue indefinitely. As Jeffrey Simpson noted, 'The courts aren't through with the abortion issue. Several of the judges in the majority sidestepped the issue of when life begins' (1988). And sure enough, only one year after this landmark ruling, the courts were once again asked to decide whose rights and subjectivity should prevail, only this time two new political subjects were introduced: the fetus and the potential father claiming ownership/control over it.

Potential Fathers, Fetuses, and the Erosion of Women's Autonomy

The fragile feminist frame of governance within which the majority of the Supreme Court of Canada attempted to locate the abortion issue did not remain unchallenged for long. In quashing section 251 of the *Criminal Code,* the court placed abortion in a legal vacuum. Abortion as a social and medical activity existed within an absence of regulation, and not as a fully articulated, legally endorsed right. However, it was not in a social vacuum and various actors were motivated to force the governmental issue.

A long-time opponent of Henry Morgentaler, Joe Borowski, a prairie activist, politician, and ardent anti-abortionist, spent much of the early 1980s in provincial and federal courts also, somewhat ironically, fighting section 251. However, his claim was that any use of federal funds for the establishment or maintenance of abortion facilities was illegal. Aided by the *Charter* as much as the Morgentaler ruling, finally, in 1983 Borowski made his claim to the Saskatchewan Court of Queen's Bench that the fetus be considered a 'legal person' under the scope of the word *everyone* used in section 15 of the *Canadian Charter of Rights and Freedoms* (the equality provision). That court held in October 1983 that it was within Parliament's jurisdiction to make decisions on the extension of legal rights of living beings to those of the 'unborn' (*Borowski* 1983). The Saskatchewan Court of Appeal upheld the dismissal, but found that that it was not the intention of section 7 or 15 of the *Charter* to protect the rights of a fetus to life (*Borowski* 1987). Finally, Borowski got his day in the highest court in the land; however, because section 251 of the *Criminal Code* had been declared of no force and effect because of the previous year's *Morgentaler* decision, once again the Supreme Court was able to dodge the issue of fetal subjectivity, dismissing Borowski's case as moot.[3]

Many interpreted the decisions in *Morgentaler* and *Borowski* as the Supreme Court's challenge to Parliament to draft legislation that would not only accord with the *Charter,* but would bring into the arena of public debate and fix, in governmental terms, the loaded issues of when life begins and what the state's interest in the fetus should be, if any. Prime Minister Mulroney complied within days of the *Borowski* decision, stating that the Conservatives would bring in new abortion policy. Divisions in the caucus over the issue led the prime minister to commit to a free vote on abortion in the House of Commons. Various policy options were 'floated' by the members of Parliament, but were all defeated, from the most restrictive to the most liberal (Pal 1991, 284–6). The divisiveness of the issue was clear. However, while the legislative branch of the state was reluctant to touch such a politically fraught issue, the courts were once again being required to act. In the summer of 1989, dubbed the 'abortion summer' by Leslie A. Pal (1991, 287), three different legal actions were brought in Winnipeg, Toronto, and Montreal, where men were requesting court orders to prevent their ex-partners from procuring abortions. This trio of cases, and the public discourse that surrounds them, constitute our second event in the development of Canada's governance strategy on reproductive biotechnologies.[4]

Following the *Morgentaler* precedent, the Winnipeg case was discussed immediately and Toby Hirsch had her abortion the same day. Justice Hirschfield noted, 'I am not convinced medically that an eight-week-old fetus is in fact a distinct human being' and suggested the overwhelming consideration was that a pregnant woman 'has an absolute right ... to the control of her body' (Weston 1989). In Toronto, however, the lower courts granted and upheld the injunction sought by Gregory Murphy prohibiting his former girlfriend, Barbara Dodd, from obtaining an abortion, and sending the case to the Ontario High Court of Justice (*Murphy* 1989a). The issue immediately became a lightning rod in the public sphere for both sides of the abortion debate. Morgentaler appeared in the news promising to provide Dodd with an abortion free of charge, at any time. Finally, the Ontario Court threw out the injunction on 11 July, suggesting that it had been obtained on false hearsay about the threat to Dodd's health. Later that day, national television broadcast images of Dodd being escorted to the Morgentaler clinic for an abortion accompanied by a crowd of pro-choice activists.

It was not over yet. Only one week after the procedure, Dodd was back in the news, reunited with her boyfriend and declaring that she regretted her decision and had been manipulated by 'pro-abortionists' (e.g., Armstrong 1989; *Edmonton Journal* 1989). Dodd became a temporary 'poster girl' for the anti-abortion movement, because of her very public conversion, and because her middle-class, teacher boyfriend appeared to be saving her from a sexually suspect life. Before meeting Murphy, Dodd had worked as a stripper, and she admitted to having had two previous abortions, in addition to two children, all before the age of twenty-one. The press revelled in presenting her as a less-than-intelligent publicity-seeking opportunist. Moreover, it was revealed that there was at least one other man who may have fathered the fetus, calling paternity into question in the case. Regardless of her past, Dodd accused members of the Ontario Coalition of Abortion Clinics of brainwashing her and insisted that the question of reproduction was not about a woman's right to choose but was a moral responsibility to the family and good citizenship. As she said in a *Maclean's* interview, 'I do not think that women should accept abortion. Especially, a woman should not decide without the father' (in Walmsley 1989, 18). Murphy was not the brooding, controlling egomaniac that she had described to the courts; he was merely her superior in class: 'He knows a lot of words and he teaches me about business, politicians and the world' (19). Further, his reasons for going to court were not about power and control: 'He loves me,' claimed Dodd, 'He wants to marry me now. That is a rare man' (18).

Echoing the Testifying ritual of the handmaids, Dodd's very public recanting placed anti-abortion discourse back in the public eye, after the setbacks that movement had received as a result of the *Morgentaler* and *Borowski* decisions. Rejuvenated by Dodd's about-face, opponents of abortion next turned their sights on the·case in Montreal between Chantale Daigle and her ex-partner, Jean-Guy Tremblay.

The Quebec Superior Court granted an injunction requested by Tremblay on 17 July 1989 (*Tremblay* 1989a) to prevent Daigle from having an abortion. Mr Justice Viens ruled that Quebec's *Charter of Rights* used the term *humans* instead of *persons* in the provisions that protected security, integrity, and personal freedom, as well as those preventing peril. Furthermore, the judge deployed a language not of *fetus* but of *a child conceived but not yet born* (para. 81, authors' trans.). He distinguished the federal *Charter*'s language of *everyone* from the Quebec *Charter*'s language of *every human being* and determined that if the Quebec legislature had meant its laws to apply after birth then it would not have implicated all human beings. Further, the maturity of the fetus was an irrelevant consideration as the fetus's right to life was absolute and began at conception (paras. 82–9). The judge asked, indignantly, with reference to Tremblay, 'Who would contest the right, moreover the obligation of a human being to want to protect his progeny and another human being?' (para. 97, authors' trans.).

Unlike Dodd, Chantale Daigle studiously avoided the media's gaze but was defiant in her right to have an abortion. Daigle quickly appealed to the Quebec Court of Appeal, where she received treatment very similar to that from the lower court. Mr Justice Bernier agreed with his lower court brethren that the fetus was a human being and insisted that 'it is his [the father's] child as much as the mother's, not more, not less' (*Tremblay* 1989b, para. 11, authors' trans.). Finally, the judge argued that to allow an abortion to take place would not only suggest that individual rights should triumph over those of the family but would irrevocably deny the fetus any rights at all. 'Allowing for an abortion against the wishes of the father would be akin to sanctioning abortion regardless of the public good. It would be to deny any judicial interest in the father, who along with the mother, is responsible for conception; it would treat the child conceived but not yet born as a non-entity' (para 13, authors' trans.). When Mr Justice LeBel (who was then a member of the Quebec Court of Appeal and eventually to be a member of the Supreme Court of

Canada) weighed the rights to life of the fetus versus the 'inconvenience' and 'difficulties' that would face Chantale Daigle, there was no contest in his mind (para. 112, authors' trans.).

The comments of the justices of the two majority decisions in the Quebec courts contrast dramatically with the language in the Supreme Court of Canada's decision in *Morgentaler.* The pregnant woman's body is seemingly effaced, or at a minimum it becomes a resource in the service of an undefined 'public good,' manifesting what Eileen V. Fegan (1996) critiques as an anti-feminist 'maternal ideology.' The pregnant woman's subjectivity is now challenged by the rights of two other subjects: the fetus and the father. Mother and father are seen to have equal ownership interests in the unborn child and therefore the issue of abortion becomes one of balancing competing, equal interests, in the service of the public good. The bodily sovereignty of a woman, her conscience, and her right to abort a fetus are swept away. Not only did the Quebec Court of Appeal reframe the language of the abortion debate to favour families and fetuses, it also acted strategically, waiting to release its ruling until the last day on which most hospitals would agree to perform an abortion, the twentieth week of Daigle's pregnancy. Undeterred, Daigle appealed to the Supreme Court of Canada, which agreed to hear her case immediately, given the urgency of her situation.

Chantale Daigle had become a somewhat unwilling 'national symbol.' 'Prochoice activists adopted her as a heroine, a victim of sexual oppression, a woman who had been denied a fundamental right. Anti-abortion activists painted her as a villain and found in the Quebec courts' rulings the first sign of hope since the Supreme Court of Canada struck down the federal abortion law in 1988' (*Chatelaine* 1990, 40). All eyes thus turned to Ottawa, as the decision on a woman's right to abort over a father's right to claim custody of a fetus was to be decided by the highest court. What no one knew at the time was that, on the same day that the hearings began, Daigle had crossed the border to have an abortion at a Boston clinic, given that Canadian hospitals would no longer conduct the procedure. She kept the secret until the final day of the hearing, three weeks later when the court was preparing its ruling. By that time, Daigle would have been entering the third trimester of her pregnancy, at which point an abortion would have been a much more complicated and physically difficult procedure, if she could even have found a doctor to perform it. According to media reports, her lawyer was so angry that he considered quitting on the spot, but relented and announced his client's actions in defiance of the law to the judges.[5]

Despite the revelation of Daigle's action, the court decided to continue with the proceedings and, by the end of the day, overturned the Quebec court's injunction unanimously, after deliberating for less than fifteen minutes. Further, Daigle would not be charged with contempt of court for having had the abortion, despite the outstanding court order prohibiting it. Taking direct issue with the Quebec courts' ruling of the fetus as a 'juridical person,' and repeatedly referring to Tremblay as only a 'potential father,' a unanimous Supreme Court stated categorically, 'The articles of the *Civil Code* referred to by the respondent do not generally recognize that a foetus is a juridical person. A foetus is treated as a person only where it is necessary to do so in order to protect its interests after it is born' (*Tremblay* 1989c, para. 61). The court equally made short shrift of the potential father's rights: 'No court in Quebec or elsewhere has ever accepted the argument that a father's interest in a foetus which he helped create could support a right to veto a woman's decision in respect of the foetus she is carrying' (para. 79).

While the court roundly denied the subjectivity of the fetus and the father in relation to the autonomy of the mother, it was adept in its avoidance of defining the moral or scientific status of the fetus.

> The Court is not required to enter the philosophical and theological debates about whether or not a foetus is a person, but, rather, to answer the legal question of whether the Quebec legislature has accorded the foetus personhood. Metaphysical arguments may be relevant but they are not the primary focus of inquiry. Nor are scientific arguments about the biological status of a foetus determinative in our inquiry. The task of properly classifying a foetus in law and in science are different pursuits. Ascribing personhood to a foetus in law is a fundamentally normative task. It results in the recognition of rights and duties – a matter which falls outside the concerns of scientific classification. In short, this Court's task is a legal one. Decisions based upon broad social, political, moral and economic choices are more appropriately left to the legislature. (*Tremblay* 1989c, para. 38)

With this, as it had in *Borowski* and *Morgentaler*, the court, in practical but also in governmental terms, returned the problem of fetal rights to the federal government, complete with the judicial recognition that there was a place for the legislature in the governance of reproduction.

The press, of course, had a field day with the fact that two very similar cases had thrown the most controversial issue in recent Canadian political history into high relief. The cover of *Maclean's* magazine blazed,

'Abortion on Trial' and featured a smiling Dodd. Journalist D'Arcy Jenish correctly noted that 'with the battle over abortion focused increasingly on individual women and their private dilemmas, the emotional pitch of the rhetoric between prochoice and anti-abortion activists has escalated' (Jenish 1989, 14–15). The perception was that the *Dodd* case and the decisions out of Quebec in *Daigle* had swung the momentum back towards the anti-abortion movement after the decision in *Morgentaler* had come down in favour of a pro-choice position. A February 1989 poll by Gallup Canada revealed that 27 per cent of Canadians believed abortion should be legal under any circumstances, almost 60 per cent favoured abortion with some restrictions, and only 13 per cent were categorically opposed to abortion (16). One journalist noted that the cases have 'sparked a burning ethical, moral and legal debate on an issue already raging out of control in the mainly lawless environment that surrounds abortion in Canada' (Weston 1989). The *Montreal Gazette* suggested, 'Fools jump in where angels fear to tread, and in this area legislators are clearly afraid of acting like fools. But they now have a duty to legislate where competing rights exist, and the three recent Canadian cases all point to a crying need for such legislation' (Paterson 1989). This call echoed repeatedly throughout the press (e.g., Beltrame 1989; Laing 1989). All of this then put an enormous amount of public pressure on the Conservative government to act, to reassert a place for the state in the regulation of abortion. Even abortion's opponents realized this and practically taunted the prime minister to resolve the situation that the courts had created by dismantling the existing law. As the spokesperson for Campaign for Life told the media immediately after the ruling, 'Brian Mulroney was trying to get the Supreme Court to do his dirty work so he could come out smelling like a rose. Now the court is going to heave the ball back into his gut, and he will finally have to face up to the political heat' (Wallace and van Dusen 1989, 14).

Bill C-43, an *Act Respecting Abortion*, was introduced by Doug Lewis, minister of justice and attorney general of Canada on 3 November 1989 and went to second reading on 7 November 1989. While the government was bound to a free vote, the Cabinet would support the bill and therefore it needed only 148 votes to pass (Pal 1991, 290–1). The legislation sought to return abortion back under the watchful gaze of the medical establishment and the legislature. Bill C-43 offered free and accessible abortions in all Canadian hospitals, provided that the woman's doctor deemed the procedure medically necessary for her health or life. Failure to prove that necessity could result in a prison sentence for the doctor of up to two years.

Abortion opponents were outraged that the legislation said nothing about the rights of the unborn and vowed to press charges against any doctor who performed an abortion on a woman whose physical health or life were not clearly at risk. Doctors were unhappy about the risk of criminal charges and, confronted with this potential threat, some very publicly announced that they would cease to perform abortions if the law were passed (Pal 1991, 292). Meanwhile abortion activists decried a law that once again made women's bodies subject to medical surveillance and supervision and would inevitably result, they suggested, in restricted access for many women. What was clear was that, as a governance strategy, the Conservative government was once again seeking a balance between the rights of the woman and the fetus; the medical profession was being charged with deciding and applying the criteria of the decision-making under the guise of health; and women were not being empowered to make decisions in relation to their own bodies. However, the prime minister was clear at the close of the debate on Bill C-43: 'It would be this Bill or nothing' (294).

The bill came back for third reading on 22 May 1990, by which time pro-choice advocate and Canada's first woman justice minister, Kim Campbell, was responsible for the bill. Campbell reiterated, 'I make no secret of the fact that it is a compromise ... From my perspective it is as far as I'm prepared to go. I can support this bill, but if it doesn't pass, it will not be possible to find another one that would have any chance of passing' (in Bolan 1990). With some members absent, Bill C-43 passed third reading in the House of Commons by a slim margin of nine votes, 140–131, and moved on to the Senate (Pal 1991, 295). With the non-elected Senate usually rubber-stamping legislation passed in the Commons, many were shocked when its free vote resulted in a tie. Under Senate rules, the law was defeated. Both Mulroney and Campbell immediately announced that there would be no further attempt by the Conservative government to legislate on abortion.

To summarize, at the hands of the courts, newly empowered with the *Canadian Charter of Rights and Freedoms*, women's rights over their bodies prevailed narrowly over fetuses, fathers, and the public interest. The legislature sought to diminish those rights in favour of increased rights for the state, the medical profession, and the fetus. The legislation would have once again denied women's decision-making autonomy and returned the power of choice to the scientific apparatus. The legislative branch of the state had acted, as demanded by citizens and suggested by the courts, and yet, that action had failed, in democratic terms, allowing the state to remove itself from the centre of the debate. Conception and reproduction were deemed ungovernable objects as long as they were scientifically and symbolically affiliated with abortion. What was required for the government to reassert control over the regulation of reproduction was a shift in the discursive terrain. Biotechnological developments were to aid greatly in this shift of language and hence governmental object. Tellingly, in the *Maclean's* cover story on abortion, the final story was one focusing on reproductive technologies, suggesting that 'technology alters the old rules' (Laver 1989, 20). The federal government turned its regulatory gaze towards new embryonic technologies, which had widespread applications for fertility, as well as other health concerns. There was a boom promised in biotechnological capital, and the development of a whole new private sector in genetic therapies. Scientifically and economically speaking, at least, embryos were emerging as a potential subject of reproductive politics to replace women. In October 1989, two months after the *Tremblay* decision, and one month before Bill C-43 was tabled, Prime Minister Mulroney announced the $25 million Royal Commission on New Reproductive Technologies. A new object of governance was being constructed, and it was not women's bodies or even their reproductive capacities, but rather, in a social science fictional turn, became the array of amazing new reproductive technologies on the horizon.

The Royal Commission on New Reproductive Technologies: Redefining Reproduction

The appointment of a genetic researcher to head the Royal Commission on New Reproductive Technologies, launched at the height of Canada's deeply fraught abortion debate, was a clear sign of the government's shifting tactics, and thus forms the third event in the development of Canada's governance of reproductive biotechnologies. After its disempowerment in the context of abortion, the governmental gaze became biogovernmental and, as a result, the object of regulation was no longer the woman as subject, but the embryo as biosubject. This change refigured the debate in scientific and medical terms. It eliminated the spectre of the discourse of women's rights; it permitted the understanding of the embryo in terms of biovalue, to borrow Catherine Waldby's rich notion (2000). She suggests that biovalue 'specifies ways in which technics can intensify and multiply force and forms of vitality by ordering it as an economy, a calculable and hierarchical system of value' (33). By reconceptualizing the debate around the embryo and its biovalue, it enabled a discussion of the public good of biotechnology without necessarily evoking the moral talk of sexual reproduction. Figured in biotechnological terms, the embryo was a genetically valuable resource rather than a person, and because its value was in its separation from the woman's body, women were reduced to the role of resource producers. The embryo was a being of the future, less a potential or actual person – as the fetus had been in the abortion debates – than a technology in the pursuit of a genetically enhanced human, and thus, social body.

The membership of the Royal Commission on New Reproductive Technologies (RCNRT) was not limited to doctors and researchers, but included lawyers, academics, and activists from both sides of the abortion issue. Patricia Baird was the geneticist from Vancouver serving as the committee's chair and C. Bruce Hatfield was a doctor specializing in internal medicine, from Calgary. Martin Hébert was a lawyer from Montreal, while Grace Marion Jantzen was a lecturer on religion from London, England. Bringing with her an already recognizable public profile was Maureen McTeer, an Ottawa lawyer, pro-choice conservative, and wife of former prime minister Joe Clark. McTeer had been integral to the lobby effort that called for the Royal Commission in the first place. With a nod to the anti-abortion activists, the government appointed Suzanne Rozell Scorsone, the director of the Office of the Catholic Family Life, Archdiocese of Toronto, while for balance it also named Louise Vandelac, a professor of sociology from Montreal and a declared feminist, presumably to offer a more secular humanist point of view. Even with Dr Scorsone on the panel, many anti-abortion advocates saw the RCNRT as a haven for pro-choice feminists. They protested the appointment of McTeer, who was a self-proclaimed feminist and was vocally pro-choice (van Dusen 1989, 16). Despite early public protests that abortion would infiltrate and shape the nature of the debate, the commission insisted that it had not been formed to deal with the political and ethical question of abortion, but reiterated its formal mandate to 'inquire into and report on current and potential medical and scientific developments related to new reproductive technologies, considering in particular their social, ethical, health, research, legal and economic implications and the public interest, and recommend what policies and safeguards should be applied' (Order in Council 1989).

The official news release announcing the RCNRT emphasized infertility therapies and genetic research, staying silent on the issue of abortion.

Hence a new relationship was being forged between genetic and reproductive technologies, one that framed the issue as one of medicine and science, advocating a strong public interest, and implying that there were dangers inherent to some of the technological developments if left unregulated. With the composition of the commission mired in controversy from the outset, and a national political leadership still smarting from its failures to manage abortion in both the courts and Parliament, the government seemed to be asserting a rational, scientific viewpoint in and through the commission. This shift would have the eventual effect of moving the issue of reproductive technologies from the unstable terrain of women's rights back to the more solid governmental ground of medical surveillance and state regulation according to the dictates of responsible parenthood and the public good. This shift, however, was neither easy nor without incident.

As a public body charged with gathering opinions, not merely from experts, but also from the general public, the site of the production of discourse around the commission expanded from the legal arena to include the domain of public opinion and the media. The rocky start of the RCNRT only got rougher as commissioners jostled for position within the commission itself as well as in the public eye. Further, it became evident that the scope of the task with which it was charged was huge – the commission was to examine more than fifty separate new reproductive technologies. It was to commission research on all issues, as well as to travel around the country meeting with citizens and experts. The commission's original date to report was October 1991, a date that was extended twice and ultimately moved to November 1993. However, the 'controversy-plagued Royal Commission on New Reproductive Technologies' ultimately broke down in December 1991 (McKeague 1991).

Reports had been leaked to the press of infighting among the commissioners allegedly due to Baird's autocratic management style, which was resulting in her making unilateral decisions about what research should be commissioned. Maureen McTeer, Martin Hébert, and Louise Vandelac were vocal in the press about their concerns. Eventually they were joined by Bruce Hatfield, and the discontented four requested a meeting with the clerk of the Privy Council Office to resolve the problems. Although the meeting took place, a solution was not forthcoming. The government responded in August 1990 by appointing two more commissioners – law professor Bartha Maria Knoppers, and former medical administrator Susan E.M. McCutcheon – both generally seen to be 'pro-technology' and intended to counter the weight of Vandelac, McTeer, and Hébert.

The second response of the government was to issue an unprecedented second Order in Council granting or confirming Baird's exclusive decision-making authority.

Under those conditions, the conflicts inevitably continued, and on 2 December 1991, McTeer, Hatfield, Hébert and Vandelac filed a suit against Baird and the federal government, critical of the research program and the structure that placed them outside of the decision-making process. They suggested that the commission was not being conducted in a manner consistent with *The Public Inquiries Act*. They were further concerned that ethical and social issues were being downplayed in relation to scientific concerns and wanted resources to prepare a dissenting report, if necessary. On 6 December Parliament adjourned, and, a mere four days after commencing their action, the four 'renegade' commissioners were fired. As a result, their suit was dismissed because they no longer had standing to make their claim.

The dramatic action of firing the dissenters resulted in calls in the press for the resignation of the chair of the commission and the reinstatement of the four. The National Action Committee on the Status of Women (NAC) and the coalition that originally lobbied for the commission called the government's actions 'extraordinary and outrageous' (in McKeague 1991). But what was becoming apparent, as Paul McKeague of the *Windsor Star* correctly noted was that 'the commission's wrangles have overshadowed the issues it was set up to study.' There was a general air of resignation in the press that bioethicist Bernard Dicken's 1989 caution had come true: 'If it [the commission] becomes a battleground between warring ideologies, it will be lost and facing what so many other royal commissions have faced when their work was done – enduring inconsequence' (in Hurst 1991). Thus, the government's technique to bring reproduction under the rubric of a broad governmental program around technology rather than subjectivity was undermined before the commission had even issued its final report.

Nor did the controversy end there. Two months after the firings, NAC released a statement to the media claiming that lucrative contracts were being given by the commission to consultants with deep ties to the pharmaceutical industry. Calling the actions of the commission 'scandalous,' the indictment by NAC also included charges that commission members had altered a report on a meeting of the World Health Organization to insert a pro-medical bias into its recommendations, in keeping with what it claimed was the commission's hidden agenda (*Edmonton Journal* 1992). The largest feminist lobby group in Canada insisted that the commission was clearly privileging 'the professional and commercial interests of a small number of scientists, doctors and medical administrators linked hand in surgical glove to a massive corporate pharmaceutical sector' over the rights of women (*Vancouver Sun* 1992). They urged other women's groups to boycott further participation in consultations with the RCNRT (Cameron 1992).

Vandelac and McTeer, angry with their treatment at the hands of Baird and the Conservative government, also continued to be publicly vocal. Vandelac expressed concerns that the commission promoted a 'narrow medical perspective' and voiced her doubts about the final report a month before it was made public (Jeffs 1993). McTeer, by far the most vocal of the detractors, gave interviews across the country, cautioning against the hasty adoption of some new technologies. 'How did it come to happen,' McTeer asked rhetorically, 'that in a society as highly educated as ours, we transferred the animal kingdom's technology into women without a blink?' (in Wigod 1993). Finally, in November 1993, the Royal Commission on New Reproductive Technologies submitted its report, two years late, $3 million over budget, and having generated fifteen volumes of research.[6]

Entitled *Proceed with Care: The Final Report of the Royal Commission on New Reproductive Technologies*, it ran 1,300 pages and made 293 recommendations. The first category of recommendations called for *Criminal Code* prohibitions on selling human eggs, sperm, zygotes, or fetal tissue; any commercial activity around surrogacy arrangements; using embryos in cloning-related research; creating human/animal hybrids; fertilizing eggs from female fetuses for implantation; and the interference with the physical autonomy of a pregnant woman. The second category of recommendations proposed the creation of the National Reproductive Technologies Commission, a body that would oversee and monitor reproductive technologies and practices, including licensing. It would have six areas of regulatory responsibility: sperm collection, storage and distribution with assisted insemination services; assisted conception services; prenatal diagnosis; human zygote research; the provision of human fetal tissue for research or other authorized purposes; and the compilation and evaluation of data on infertility, infertility research, and options for preventing or reducing infertility. More active roles were recommended for certain federal departments as well, such as Health, Environment, and Human Resources.

There was clearly an attempt by the commission to divorce market forces from choices around reproduction, and at the same time a strong endorsement of the role of the family in a healthy Canada. There were concerns in the report about the discrepancies in available services and practices across the provinces. Most significantly, from a biogovernmental point of view, was the call for a regulatory commission – similar to the ways in which the Canadian Radio-Television and Telecommunications Commission regulates telecommunication and broadcasting – to provide governmental oversight of infertility generally, and the intersection between reproduction and biotechnologies more specifically. The clear message was that the risks of these technologies, and in particular, their commercial potential, needed to be managed by a state-authorized agency.

The recommendations that caught the media's eye included the banning of surrogate motherhood, sex selection clinics, and the sale of human sperm, embryos, and fetal tissues, as well as the call to make in vitro fertilization techniques available on a more widespread basis, despite their problems (e.g., Page 1993). Presented with a report that addressed technologies and practices contemplated only in science fiction, such as human cloning, creating human-animal chimeras, and placing human embryos in other species, journalists could not resist references to *Brave New World* and *The Handmaid's Tale* in their commentaries. More conservative media suggested that the RCNRT had struck a balance, managing to avoid the risks that Canada would become like the United States, where battles between surrogate mothers and adoptive parents over contracted babies were being waged (e.g., *Ottawa Citizen* 1993). The *Globe and Mail*, however, claimed that the commission had demonstrated a bias against individual rights and an 'equally wrongheaded prejudice against commerce' (*Globe and Mail* 1993). The editorial board of the *Vancouver Sun* agreed, denouncing the report for its 'Big Sister' approach and lack of commitment to sound scientific values and economic progress. Calling the commission a 'royal busybody,' it criticized the commission's 'repugnance for commercialization' and shortsightedness over embryonic research: 'Only through research can science make the discoveries that benefit humanity' (*Vancouver Sun* 1993). Legal analysts, too, criticized the recommendations (e.g., Healy 1995).

Despite the controversy, the Royal Commission on New Reproductive Technologies as a biogovernmental event, signalled the clear beginnings of a shift in the governmental problematic, reworked from a framework of rights (women's, fetus's, father's) and into one of responsibility, in which women's bodily integrity was removed from the battlefield of the abortion debates and immersed in a different kind of power struggle. In

a responsibility frame, the state is authorized to manage the reproductive body as a passive object, while simultaneously ascribing utilitarian
biosubjectivity to the embryo. In doing so, it elevates the role of medical
technologies in providing the state with the techniques to displace
women from the centre of reproduction and replace them with the abstracted concern over fertility. This is the shift evident in the contrast
between Atwood's two reproductive dystopias. In *The Handmaid's Tale*,
the state cannot fully realize control over women, since women's bodies
remain necessary, and thus the compliance of women as subjects must be
won. In *Oryx and Crake*, however, written in the wake of the scientific explosion in reproductive technologies, Atwood dramatizes how these
technologies can give the medical and research communities broad access and control, leading to a situation in which women's bodies cease to
be a necessary component in reproduction. Within the emerging
biogovernmental frame that we argue their RCNRT represents, women
are interpellated as responsible, with the male partners, for the maintenance of a healthy nation, through the vehicle of the family.

Gerlach (2004) defines responsibilization as a key component in
biogovernance, one that transforms the risk of a biotechnological society
from state intrusion into personal rights, to individual responsibility for
social improvement through the appropriate deployment of biotechnological resources. The responsibility frame, identified by Piet Strydom as
a monumental discursive shift, helps to usher in what he calls the 'civilization of the gene,' as it bridges legal and scientific regimes to resolve
what has become known as the 'problem of nature' (1999, 33). Infertility
is characterized as the 'problem of nature' to be fixed, in the very terms
of reference of the commission and particularly in the responses to its
recommendations. Interestingly, in *Oryx and Crake*, nature, or the biological imperative to mate, is aligned with both erotics and the imagination, and once it has been rationalized and controlled, is lost and
mourned. At the heart of responsibilization, then, is a double bind, in
which science has both opened society to new levels of risk and positioned itself as the only social force able to resolve potential problems
– problems redefined as coming from nature and not from the technologies themselves.

The emergence of the responsibility frame has been felt very keenly in
the arena of sexual and reproductive politics, insofar as they contend
directly with the limits of bodily integrity, or of private rights over public
responsibilities, and are arguably structured by a latent belief that medical intervention can lead to the perfection of human life, 'doomed by

hope,' as the narrator of *Oryx and Crake* put it. Canadian governance and its corresponding legislative agenda took a sharp detour away from contraception and abortion and into an uneasy relationship with conception and genetic biotechnologies. It centred attention on the isolated, autonomous embryo in the laboratory Petri dish, rather than on women's bodies or even the fetus within the woman's body. The RCNRT was both productive of, and emerged within, this evolving framework of responsibilization, charged with mapping the place of reproductive technologies in Canadian society. This tentative discursive frame would solidify over the course of the fourth event we identify in Canada's quest to govern reproductive biotechnology: legislation on reproductive and genetic technologies.

Biogovernance Settled: The *Assisted Human Reproduction Act*

The attempt by the RCNRT to create a contingent biovalue of the embryo as long as it remained within the laboratory and outside of women's bodies was challenged on multiple fronts by critics in the media, family planning groups, and the biotechnology industries. The newfound victims of the commission's short-sightedness were families. Under the headline 'Banned Parenthood,' *Maclean's* ran a story of a young mother who had agreed to act as a surrogate for $13,000 in order to try to save the family farm (Nichols and Driedger 1993, 52). CBC interviewed a woman who, in an advertisement, offered to be a surrogate for $12,000, stating, 'I'm useful. I could give life' (Prime Time News 1994). Thus, increasingly, media reports in particular publicly defined the issue of reproductive technologies according to a political-economic logic of labour rather than sexual rights, and the free choice to reproduce, rather than the option not to. This refocused the debate to one of fertility, rather than limiting conception for the sexually active woman. Feminist organizations and activists could then be coded as the enemies of families. The abortion debate with its focus on individual rights – women or fetuses – began to sound dated as the issue of reproductive technologies was reframed around the family and the need to give hope to couples trying to conceive. A charismatic technological future beckoned once again. The embryo was carefully extracted from claims to either moral or juridical status, understood entirely as a biosubject – given its value only in the laboratory through techniques of medical science. It was an entity produced of, and representing, biovalue, despite limitations placed on its participation in the marketplace.

Proceed with Care had clearly lobbed the governance ball, as the courts had done previously, back into the hands of the government. It was difficult to see how the government could avoid legislating on at least some of the issues surrounding new reproductive technologies. What ensued, however, was a ten-year struggle to define the role of reproductive technologies as objects of governance without having to resolve the issue of reproduction. The volatile and uneven path towards the current legislation, Bill C-6, the *Assisted Human Reproduction Act*, is our fourth and final event in the story of the biogovernance of reproduction in Canada.

In July 1995, Health Minister Diane Marleau had urged a voluntary moratorium on nine controversial reproductive technologies, including sex-selection techniques, commercial contracts for surrogate motherhood, and trade in human reproductive materials. However, practitioners in the fertility industry continued their work despite the government plea. Their rationale was that they were engaged first in the noble work of helping infertile couples have children, and second in conducting research that would produce lines of gene therapies to reduce and eliminate debilitating diseases and conditions. Against such powerful symbolic, scientific, and economic claims, clearly legislation would be needed for the state to assert some control over reproductive biotechnologies.

Bill C-47, the *Human Reproductive and Genetic Technologies Act*, was a hasty piece of legislation introduced to the House of Commons by Health Minister David Dingwall on 14 June 1996. It did not proceed past the House, however, because a federal election was called in the spring of 1997 and Parliament was dissolved. However, it did play an important role in marking out what the points of contestation would be in the future debates on subsequent attempts to legislate.

The preamble to Bill C-47 framed the issue of reproductive biotechnologies as one of risk management. 'The Parliament of Canada is gravely concerned about the significant threat to human dignity, the risks to human health and safety, both known and unknown, and other serious social and ethical issues posed by certain reproductive and genetic technologies.' Further, the objectives of the legislation provided in section 3 were largely paternalistic:

(a)   to protect the health and safety of Canadians in the use of human reproductive materials for assisted reproduction, other medical procedures and medical research;

(b)   to ensure the appropriate treatment of human reproductive materials outside the body in recognition of their potential to form human life; and

(c)   to protect the dignity of all persons, in particular children and women, in relation to uses of human reproductive materials.

Through this language, only 'persons' are protected, knowing that the courts had already established that fetuses (and certainly not embryos) were not persons at law; reproductive technologies remain a danger against which the Canadian population must be protected; and the embryo as biogovernmental subject is categorized in object terms with other forms of 'reproductive materials' such as cells, sperm, or ova. A clear piece of legislative social science fiction, Bill C-47 banned sex selection; commercial surrogacy; buying and selling of ova, sperm, embryos, or fetuses; germ-line genetic alteration; ectogenesis; cloning of human embryos; human-animal hybrids; and the collection of ova and sperm from fetuses and cadavers for reproductive purposes. As one journalist remarked wittily in the *Calgary Herald*, 'If there is a Bond-movie villain in a luxury bunker carved out of a mountain deep in the Rockies creating a human-animal hybrid to take over the world, we can rest assured he will soon be officially breaking Canadian law' (Trigueiro 1996).

The legislation was roundly criticized, particularly by the medical community, for focusing on a list of prohibitions, rather than on an articulation of what was possible and permissible. Doctors were not happy with the prospect of fines up to $500,000 and jail terms of up to ten years for violating the prohibitions of the legislation. They critiqued the ban on nominal compensation for sperm donors and were specifically concerned with the social science fictional character of the legislation. Medical research associations were concerned about bans on procedures that were currently scientifically impossible and 'in the realm of science fiction' (in Moysa 1996). They felt that potentially life-saving or promoting techniques should not be prohibited in advance of their development.

Former members of the RCNRT commented on the legislation, with Louise Vandelac arguing the legislation did not go far enough to protect women (Fidelman 1996); Maureen McTeer calling for a licensing body and the separation of reproductive and genetic technologies (McTeer 1996); and Patricia Baird suggesting that the regulatory body recommended by the commission was overdue and badly needed (Baird 1996). However, specifically feminist critiques focusing on the rights of women and critiques of the biotechnology industry did not resonate as they had a decade earlier (e.g., Fidelman 1996), in large part because the terms of the debate had changed.

From our perspective, what is interesting is the dramatic shift in frame between the abortion debates and these legislative scuffles. The spectre of abortion had been at the heart of the *Morgentaler, Daigle,* and *Tremblay* cases, and its ghost hovered continually over the Royal Commission on New Reproductive Technologies. By shifting the optic to a fusion of reproductive and genetic technologies targeted at infertility – the issue of desired children, rather than undesired pregnancies – the state and medical profession alike could fall back from far more treacherous terrain and relocate the debate to more secure ground. *Maclean's* signalled this new ordering in its cover story, 'Beyond Abortion: Advances in Science Leave an Old Debate in the Dust' (Wood 1996, 14), which concentrated on the recently introduced legislation. *Maclean's* coverage represented the issue romantically with images of newborns emerging into the welcoming arms of families that otherwise would not have been, coupled with images of scientists experimenting on embryos for the health of the population. In this way, parenthood became a strategic offshoot of benevolent scientific research, and, in order for it to be 'left in the dust,' abortion had to be characterized as a misguided attempt to personify the embryo/fetus. From a governmental perspective, the abortion debate had to be vacated in order to enable the biovalue-generating activities of the medical research community, all in the name of the family.

In February 1997, when researchers in the United Kingdom announced the cloning of a mammal from an adult female, Bill C-47 took on new importance. The birth of Dolly the sheep sent shockwaves around the globe and spurred a new debate on the scope of genetic technologies and the place of governments in their regulation (see Gerlach and Hamilton 2005). The legislation suddenly received even greater scrutiny from the public and the scientific community, alike. The calls for a regulatory body, in particular, grew louder. Critics were additionally concerned that the legislation was being 'fast-tracked' through Parliament with only two brief roundtable discussions from interested stakeholders, the most limited public hearings on new legislation possible. In effect, Dolly crystallized the opposition to a piece of social science fictional regulation (see Hamilton 2003). An editorial in the *Ottawa Citizen* claimed, 'Dolly has spawned thoughtful worries about the course of technology, but also overblown fears of a host of Frankenstein's monsters bursting forth from some hellish laboratory. The sudden furor over heretofore futuristic technologies has exposed C-47 to the sunlight, just as the bill enters the last leg of its parliamentary journey. Thank you Dolly' (1997).

At first, the Liberal government stood defiantly in support of its measures, citing the crisis brought on by Dolly as demonstrating the need for strong and decisive governmental action; however, the pending legislation became a casualty of a spring 1997 election call and the dissolution of Parliament. After the flurry of activity around Dolly, the issue seemed to fall off the government agenda. By 1999, *Maclean's* posed the question that many were wondering: 'In the age of in vitro fertilization does the state have a place in the test tubes of the nation?' (Geddes, Mcclelland, and Chisholm 1999).

With the Health Ministry struggling to keep pace with changing public attitudes as biotechnologies rapidly came onto market, a new stakeholder emerged that gave a distinct advantage to the growing support for the commercialization of reproductive and genetic technologies. In 1998, Industry Canada, the federal ministry in charge of productivity and competitiveness in a knowledge-based economy, singled out biotechnology for major growth and investment. The era of genetic commerce had arrived, supported by a report from the Canadian Biotechnology Advisory Committee (CBAC), contending that Canada could become a world leader in the field, gaining as much as 10 per cent of a $50 billion market – with the right kind of governmental support and regulation. The deputy minister of finance was reappointed to Health, and CBAC recommended that a regulatory agency be established that would report directly to Industry Canada. The *Toronto Star* commented on the shift in focus from health to commerce: 'The government has been slow to react because the field is a difficult one. Some possibilities in genetic commerce seem to lead to ethical swamps. But many of the developments may make life more pleasant for more people' (Ford 1998). With the collapse of C-47 and the realignment of biotechnology with Industry Canada, it seemed that an enfeebled government sector was bowing to the corporate interests of a burgeoning biotech industry.

In May 2001, Minister of Health Anne McLellan asked the House of Commons Standing Committee on Health to review the government's *Proposals for Legislation Governing Assisted Human Reproduction*. That committee's report was tabled in December of that year, tellingly entitled *Assisted Human Reproduction: Building Families* (see Brown, Miller-Chenier, and Norris 2003). It strongly recommended that the government prioritize the drafting of legislation, and Bill C-56, *An Act Respecting Assisted Human Reproduction*, introduced into the House of Commons on 9 May 2002, reflected the committee's approach. Dramatically less draconian than its predecessor, Bill C-56 framed the issue, not as human dignity, but as one potential of reproductive technology. Section 2(a) is illustrative of the reframing of the issues from the dangers of technology run amok, to the investment in the possibilities of genetic science, properly regulated:

> The Parliament of Canada recognizes and declares that:
> (a)   the benefits of assisted human reproductive technologies and related research for individuals and for society in general can be most effectively secured by taking appropriate measures for the protection and promotion of human health, safety, dignity and rights in the use of these technologies and in related research.

Commercial surrogacy was still banned outright (section 6); however, significantly, the act provided for the creation of an Assisted Human Reproduction Agency of Canada, which would grant licences to conduct therapeutic research on human embryos. Its dictates would be binding on the private sector, as well as on the government.

Despite more flexible legislation that addressed some of the criticisms levelled at C-47, once again, the critics lined up. Maureen McTeer was concerned that it criminalized reproductive technologies and did not aggressively enough pursue the regulatory agency (1999). Former Parliamentarian Preston Manning argued that the legislation, while a step forward, required amendments. The regulatory agency, he suggested, while a positive initiative, needed to be more accountable to Parliament. He also argued that the ethic of human life needed to be more central to the entire legislation and that an affirmation of the quality of life of disabled persons should be added (Manning 2002). The same Canadians who were opposed to abortion tried to fire up debate on the legislation's endorsement of embryonic research, provided the embryo was not created by the scientist for the purpose of research (e.g., Hamel 2002, 6; *Hamilton Spectator* 2002), and its condoning of sperm and egg donation because it denied accurate parentage information to children (Clute 2002). These arguments were pre-empted as once again Parliament was prorogued in September, and once again the government's efforts to regulate biotechnological reproduction ended.

The government tried again in December 2002, with very similar legislation, Bill C-13 also entitled *Assisted Human Reproduction Act*. Interestingly, however, this time the order of priorities to be considered in regulating reproductive technologies had changed. Children conceived through these technologies had been elevated to the first priority, followed by individuals, families, and society in general (section 2(a), (b)). Women and children, once discussed together, were decoupled, and section 2(c) specifically recognized the differential impact of these technologies on women. The legislation was largely the same as its previous incarnation, but the issue that drew much of the debate was whether or not therapeutic cloning of embryos should be permitted for stem cell research. The legislation banned outright reproductive cloning, where a cloned embryo could be implanted in a woman's womb and brought to term, as well as therapeutic cloning where a cloned embryo could be used for research and medical purposes. However, it did permit research on embryos that would otherwise be discarded from both fertility and abortion clinics.

Once again, however, in November 2003, the government of Jean Chrétien recessed Parliament, and once again Bill C-13 died on the order paper in the Senate. Finally, in early 2004, Bill C-6, the *Assisted Human Reproduction Act* was re-tabled and finally adopted at the end of March 2004. More than a decade after the Royal Commission on New Reproductive Technologies had reported, Canada had legislation governing biotechnological reproduction. The legislation bans human cloning but permits embryonic research in certain circumstances. It requires researchers to be licensed and criminalizes failure to adhere to its precepts. It continues with the long-promoted government principle of banning sex-selection, commercial surrogacy, and trade in human reproductive substances. Activities that are permitted are to be regulated through licensing regimes. The bill still has its critics, but the consensus seemed to be, contrary to the abortion debate that seemed to be from another era, that some legislation was better than no legislation at all. Throughout the long legislative process, it had always been clear that the public saw a role for the Canadian state in the regulation of the application of biotechnology to reproduction, and finally it took up that role.

The *Assisted Human Reproduction Act* received royal assent on 29 March 2004. However, it was not until January 2006 that the Conservative government of Stephen Harper finally established the federal regulatory agency Assisted Human Reproduction Canada (AHRC). It comprises fourteen members, a majority of whom are doctors, and its mandate is to protect and promote the health and safety, human dignity, and human rights of Canadians who use or are born of assisted human reproduction technologies, and to foster ethical principles in assisted human reproduction and other related matters under the act. Its Board of Directors is responsible for managing the agency, including the licensing of activities under the purview of the legislation and advising the minister on policy matters. It reports to a committee of Health Canada constituted for this purpose.

Critics quickly pointed out that a significant number of members held socially conservative viewpoints about abortion and embryonic research. An editorial in the *Globe and Mail* claimed there had been 'a stacking of the board with people of socially conservative views – views that might steer the panel away from embracing scientific advances that might help those seeking to give birth to healthy children' (2007). Other media reports and opposition politicians pointed out that no genetic researchers or fertility experts were included, neither were any representatives from advocacy groups for the infertile. Finally, there was clearly no attempt to include representation from pro-choice or women's groups. The contrast, therefore, with the membership of the Royal Commission on New Reproductive Technologies, appointed by the precursor Progressive Conservative party, is striking evidence of the wider conceptual shift that has taken place from abortion and women's rights to fertility and the production of families. The critics, however, also demonstrate the change in priorities, in that, while criticizing the conservative membership of the agency, they express that critique in terms of fertility issues, the desire to give birth to 'healthy' children, and the progress of science. Gone are concerns about women's bodies and their rights in issues of reproduction. Just as the Handmaids were reimagined as the genetically engineered Children of Crake, the legal issue of women's sexual rights has been replaced by technological debates over the biovalue of embryos.

Conclusion

With the passing of the *Assisted Human Reproduction Act*, a chapter in the debate on reproductive technologies in Canada was closed. As this review of Canada's complicated path of governance suggests, the discursive framework in which governmental action ultimately would and could be articulated and justified was gradually disconnected from issues of women's bodily integrity and subjectivity. In its place is a more health- and economy-grounded alignment with science that places its emphasis on embryos as biosubjects. As Marie Fox recognizes in the context of the British experience with the governance of reproduction, 'The increasing visibility of the embryo ... thus corresponds with an erasure of the woman by a constellation of factors, including foetal imaging techniques, representations within popular culture, Pro-Life rhetoric, and the discourses of law and science' (2000, 174). The Canadian governmental rationale, too, shifts from one of the defence of individual women's bodies to the protection of the health and well-being of the national body. From the state's perspective, treating the governmental problematic as one of re-production intimately connected to women's bodies divided the elector-ate and the government, and disabled governmental action. From the beginning, with the seeds of family planning at work in the emergence of birth control, sex without reproduction could come to be viewed as socially invaluable, not an appropriate governmental focus. Unruly sub-jects, women who might not want to reproduce, could be effaced, en-abling infertile heterosexual couples wanting to have children as the new and 'safer' biogovernmental subjects of choice. The 'family' thus replaces the 'woman' as the subject on whose behalf governmental ac-tion will be taken and firmly locates the 'problem' to be regulated as one characterized in scientific and social, not personal or embodied, terms.

In what is clearly a classic shift from discipline towards biopolitics, the governmental focus, and, as importantly, the terrain of political contesta-tion, shifts from the control of women's bodies to the management of fertility. As a result, the embryo becomes the biosubject of choice. At the same time, women's bodies have seemingly been decoupled from hu-man reproduction, rendering reproduction a social science fictional technique, with what promises to be interesting consequences, intended and unintended, negative and perhaps not so negative. The new govern-mental task is to manage the emerging economy of fertility in the public interest. Biopolitical governance does not necessarily demand Crake's future vision but does require us to understand that any sexual politics of biotechnology must be aware of the current governmental frame; exam-ine the diverse consequences that arise from this shift, in particular, a range of new biosubjects; and creatively explore the possibilities, not only the risks, of contingent embodiment.

CHAPTER FOUR

Biopatents and the Ownership of Life

Well, you're not like me. Um. I mean, you're not human. I mean, you're human, but you're not real. Uh, you're not like a real person. Like me. You're clones ... Clones. Some hag trophy wife needs new skin for a facelift or one of 'em gets sick and they need a new part; they take it from you.

McCord in Bay, *The Island* (2005)

Madame *was* afraid of us. But she was afraid of us in the same way someone might be afraid of spiders. We hadn't been ready for that. It had never occurred to us to wonder how *we* would feel, being seen like that, being the spiders.

Kathy H. in Ishiguro, *Never Let Me Go* (2005, 32)

Talking about a Mouse

This is the story of a mouse, a mouse known both as the Oncomouse and his/her/its alias, the Harvard Mouse. The first name: generic/genetic, the second: legal/proprietary; the first: evoking a frightening disease; the alias: an Ivy League institution. In this chapter, we opt for the moniker Oncomouse, because it contains the traces of the laboratory while troubling proprietary logic. It also locates the cancer-gene-carrying rodent firmly in the 'molecular optic,' where 'life is now imagined, investigated, explained, and intervened upon at a molecular level' (Novas and Rose 2000, 487). In the molecular optic, one can imagine and invent not only new ways of being, but even new beings.

The Oncomouse is such a being, a 'technobastard' in Donna Haraway's terms (1997, 78). A transgenic animal, the Oncomouse is an organism that contains a gene or genes of another animal, which have been introduced into it through artificial means. In the case of the Oncomouse, this transfer occurs at the embryonic stage – the fertilized eggs are transferred to a female 'foster mouse' and permitted to gestate naturally. The gene in question is an oncogene, which predisposes the animal to developing neoplasms, or malignant cancerous tumours. The resulting mice are then tested to see if they have retained the gene; those that have are called the 'founder mice.' Founder mice are then bred with regular mice, and so on, and so on, à la Mendel. Once the female mice with the injected gene give birth and begin nursing, they grow tumours in their breasts. Oncomice, as reliably cancerous beings, are obviously very useful to researchers testing both carcinogens and cancer-treating drugs. They entered the market at $50 per (transgenic) animal, and *Fortune* magazine listed the Oncomouse on its '10 Hottest Products' list in its 5 December 1988 issue.

After Mickey, the Oncomouse may well be the most famous global mouse citizen. Financially, the Oncomouse would give its cartoon kin a run for his money, and, interestingly, both have played very significant roles in the development and expansion of intellectual property rights in North America and around the world, in the past fifteen years.[1] For Oncomouse, this is because he/she/it is also known by a number of other code-names, among them #04, 736, 866 in the United States, and #484, 723 in Canada. These are, of course, the patent application numbers assigned to Oncomouse, when he/she/it became the first mammal in the world to be patented.

On 12 April 1988 two genetic researchers, Philip Leder of Harvard Medical School and geneticist Timothy Stewart of San Francisco, received a patent from the United States Patent and Trademark Office for the processes that produce Oncomice, and for the animal, or end product, itself: the Oncomouse. In other words, all offspring of any Oncomouse would be owned by them. Furthermore, the application was not limited to mice, but encompassed any transgenic mammal containing the oncogene. The two inventors assigned their patent to the president and trustees of Harvard College, who then pursued legal and commercial opportunities, including seeking patent protection in the European Union, Japan, and Canada, as well as licensing the patent for commercial development to E.I. Du Pont de Nemours (an original sponsor of the research).

Although a genetically modified microorganism had been patented in 1980, with the assistance of the United States Supreme Court (in the well-known and much discussed *Diamond v. Chakrabarty* case), the Oncomouse was the first complete mammal, the first 'higher' or 'complex' life form, to be the object of a patent. Typically the patent offices of Western nations work in relative obscurity, outside the glare of public scrutiny. The decision to grant the patent for Oncomouse, however, triggered widespread protest in the United States, including demonstrations, lawsuits, and provocative patent applications. For example, animal rights activists and farming leaders filed a lawsuit in the summer of 1988 to stop the government from issuing any further patents for genetically engineered animals. The National Council of Churches and the Humane Society urged a moratorium on the patenting of animals. The House of Representatives rushed through a bill prohibiting the patenting of human beings. The celebrated bio-pessimist Jeremy Rifkin and Dr Stuart Newman filed a patent for a combination of a human and animal embryo to produce a chimera in order to provoke public debate and show where this kind of science could lead. It was rejected by the PTO in June 1999. The application to the European Union patent office also generated much public outcry. A coalition of twenty-nine European environmental and development groups called Patent Concern was active. A coalition led by the British Union for the Abolition of Vivisection and Compassion in World Farming (twenty-five organizations from twelve European countries) launched a campaign after the provisional patent was allowed in May 1992. Protests were organized in a number of European countries.

All this talk mattered. In the United States, the Patent Office self-imposed a five-year moratorium on patents for living organisms; the EU adopted a mandatory ban for four years. In Europe, the application for Oncomouse was originally denied, although the patent was finally granted to Harvard in its second attempt in 1992, despite ongoing public dissent. More than three hundred NGOs and green political parties mounted an ultimately unsuccessful campaign to revoke the patent.

Leder and Stewart first applied for a Canadian patent for the Oncomouse on 21 June 1985. The original patent examiner rejected a significant number of the claims, including that for the animal itself. After requests for further review, the examiner in charge of the file issued the Final Action in March 1993, permitting claims to the gene-splicing processes, but not for the organism as a whole. In September, Harvard requested a review by the commissioner for patents and a hearing before the Patent Appeal Board. The hearing took place in July 1994. The decision of Commissioner for Patents Mart Leesti was released on

4 August 1995 and held that no patent would be granted for a non-human mammal. Unlike that in the United States and Europe, however, this decision – while posing a significant break with emerging biopatenting policy around the world, and concomitantly raising significant scientific, moral, and legal issues – barely registered with the Canadian public. There were no protests; there were no cries of support for the patent commissioner; no coalitions mobilized. Unlike their American and European counterparts, everyday Canadians were not talking about the Oncomouse.

But while the media and the Canadian public were not talking about the refusal to patent Oncomouse, there were other actors in the public sphere who were. This chapter examines the 'talk about a mouse,' and the Canadian institutions implicitly authorized to decide if someone can own a genetically modified higher life form. In Canada, these bodies are the Patent Office, the Canadian Biotechnology Advisory Council, and the federal courts. Throughout, a structuring absence in these discussions is Parliament, which retains its legislative silence to this day on the issue of patenting higher life forms.

The talk about a mouse is significant because it is also talk about life itself, in that when demarcations as to who can own life are defined, those demarcations generate truths about what life is, where and how it begins and ends, and who can regulate it. As Haraway suggests, 'Because patent status reconfigures an organism as a human invention, produced by mixing labor and nature as those categories are understood in Western law and philosophy, patenting an organism is a large semiotic and practical step toward blocking nonproprietary and nontechnical meaning from many social sites – such as labs, courts, and popular venues' (1997, 82). Life as an object of governance cannot remain unchanged after all this talk about owning a cancerous mouse. Most particularly, biopatenting serves as a site of contestation, 'strongly influenced by a language of self-possession,' as Margaret Davies and Ngaire Naffine assert (2001, 156). In other words, biopatenting is about the reconstitution of life as property: a powerful social science fiction. And if life is or can be property, a rupture point is produced in the construction of the self. Biopatenting becomes a central technique in the creation of a new genetic mode of subjectivity. Explored at the powerful juncture of law and science, and not only within science fiction narratives and the history of slavery, we argue in this chapter that people, along with rodents, are biojuridical subjects, always simultaneously subjects and objects, and therefore potential property.

Biopatenting

The patenting of genetic material, or biopatenting, is constituted in the application of intellectual property rhetoric, bureaucratic regimes, and legal and scientific expertise to the elements of plant, animal, and human life. Biopatenting brings together life and law, two domains that currently enjoy significant social authority in Western societies and have an ongoing relationship. Both law and science have played key roles in the production and separation of nature and culture at the heart of biopatenting (see Pottage 1998). However, biotechnological science is now materially and symbolically troubling that simple (and simplistic) distinction. And that troubling is playing out in, and being shaped by, intellectual property law. The line between the natural and the artificial has never been so unclear, bringing the relationship between science and ethics into sharp relief. As Ari Berkowitz and Daniel J. Kevles note, DNA, as the code of life, with its near sacred symbolic status, makes it somewhat volatile as a legal commodity (2002, 75). And yet, these are the distinctions in and through which the law has structured regimes of rights and duties, and from which it has taken much of its own legitimacy within modernity. Biopatenting thus makes visible the increasing instability of a related series of modernist distinctions, dually endorsed by law and science – nature /culture, natural/artificial, human/animal, invention/discovery.

All of this is compounded by the operation of biotechnological science within a global industry of biovalue. As Waldby argues, and as explained in the previous chapter, biovalue is generated whenever living entities can be instrumentalized in a manner that renders them useful as human products, from the medical to the industrial to the agricultural. Indeed, biopatents have been one of the foremost technologies of biovaluation. But, unlike a number of other scholars (e.g., Berkowitz and Kevles 2002; Bowring 2003; Hanson 2002; Rose 2001; Seide and Stephens 2002), we do not understand biopatenting as only, or even primarily, an instrumental process of positive law, mobilized by transnational pharmaceutical and bioscience companies to commodify life. Rather, we contend that biopatenting is a site of social, cultural, economic, and political rupture – a site in which otherwise often invisible operations of biopower become visible, and a site that is conducive to forms of biosubjectivity, which exceed their commodity status.

Patent Law in Canada

In general, intellectual property is a body of statutorily created, limited property rights granted to the products of human intellectual and creative effort. It includes copyrights, trademarks, patents, industrial designs, integrated circuit topography, trade secrets, and publicity rights. Governed by the *Patent Act* in Canada, the technical elements of patents have changed very little over the years, but the subject matter to which those technical elements apply have obviously changed dramatically. In 1869, when the legislation first came into force, its drafters could not have envisaged insulin, the space shuttle, DVD players, Viagra, ultrasound machines, and the host of Starfrit appliances that emerged over the next 150 years. But the legislation has expanded to take account of all of these items, primarily through the interpretation of what constitutes an invention. Through the concept of invention, patents and patent systems have had a dynamic relationship with the development of new technologies and scientific techniques.

A patent is a monopoly property right granted to an inventor for an invention. An invention is defined in section 2 of the act as 'any new and useful art, process, machine, manufacture or composition of matter,' or any useful improvement to any of those. An invention is often contrasted with the 'mere' discovery of something, pre-existing in nature that is not patentable. Section 27(8) of the act specifies that 'no patent shall be granted for any mere scientific principle or abstract theorem.'

To receive patent protection, the invention must be new, non-obvious, useful, and fully described in the application. The requirement of newness means that it cannot be something already in active use or already known. It cannot have been made public prior to the patent application, although in Canada inventors are given a one-year grace period. The invention must be non-obvious to someone trained in the particular field in which the invention operates. The usefulness criterion tends to require some form of industrial or commercial application. Finally, both the process for its production, and the invention itself, must be described in the application in such a way that someone trained in the field could produce the invention from the description.

Unlike copyright, a patent does not vest in the act of creation or invention; rather, an application must be made to the Canadian Intellectual Property Office (CIPO), the body responsible for the administration of patents. If granted, the patent is presumed valid unless and until challenged, and there is no public policy basis for CIPO to refuse a patent. A patent may be denied only if it does not meet the statutory requirements. It gives the inventor the right to prevent others from unauthorized creation, use, or sale of the invention for twenty years from the date of filing (or seventeen if the patent was filed prior to 1989). Interestingly, a patent is a negative property right – it does not give the right to the patentee to make, use, or sell the invention, only to prevent others from doing so. For instance, there may be other legislation or public policy that prohibits or restricts the practice or sale of the invention, as in the case of weapons, drugs, and environmentally volatile products, for example.

The logic behind patent law is that the period of monopoly permits the inventor the time to exploit her or his invention free from competition, in order to recoup research investment. In this way, patent systems seek to provide incentives for invention. In exchange, the patent holder is required to make full disclosure. After the patent period, the invention reverts into the public storehouse of knowledge. As a result, patents are often framed by courts and policymakers as a bargain where the inventor receives protection and financial incentives, and the public receives useful knowledge circulating openly in the public sphere.

Patents thus operate on the assumptions of the Mertonian ideal of non-interested, non-commercial research carried out by individuals, who are part of a shared community of knowledge and values. Some critics argue, however, that patenting actually deters innovation. Creators spend much time developing, negotiating, and often litigating licensing agreements and royalties. Holders of patents use them strategically to prevent competitors from developing new products. This occurs, in part, because an overwhelming majority of patents are not, in fact, awarded to inventors, but to their employers – the large corporations that fund the research. In addition, scientists frequently wait to publish their research until they can obtain the patent, so as not to jeopardize their ability to do so. Consequently, the emphasis on patents can create a chilly climate wherein researchers are reluctant to pursue certain avenues of research because they are concerned about infringing upon someone's patent. In fact, in 2002, leading cancer researchers in the United States were charging that DuPont's troublesome licensing practices deterred them from undertaking cancer research using Oncomice.

Within the domain of property rights, patents are somewhat limited in scope and range, as well as being very technical. Nevertheless, patents have recently emerged as one of the most significant sites of intellectual property because, as Mark Hanson suggests, 'as a rhetoric of the market and property rights, patents are a rhetoric of ownership, control, and assertions of sovereignty' (2002, 172). Ownership, control, and sovereignty become very powerful terms of engagement when attached to what have traditionally been viewed as higher or complex life forms. They become the cornerstones in the structures of biopower.

In fact, biopatenting is a biopolitical site par excellence – a site where biotech corporations, courts, individuals, populations, researchers, and governments all negotiate the ways in which life as a whole, and its various elements, are subject (or not) to ownership, manipulation, and alienation as property. Truth discourses are produced, authorities emerge as speakers of those truths, governmental strategies are mapped, and a new mode of subjectivity, with its own set of techniques of the self, emerges. Interestingly, it appears that, even writing in the mid-1970s, Foucault understood the biotechnological horizon point of biopower: the genetic engineering of life itself, since '[the] excess of biopower appears when it has become technologically and politically possible for man not only to manage life but to make it proliferate, to create living matter, to build the monster' (2003, 254). Biotechnology has made it technologically possible to build the monster; patent law is making it politically possible.

Biopatenting in North America

Within the very premises of patent law – the separation of 'mere discovery' (nature) from 'invention' (culture), and the insulation of patents from politics through the absence of public policy considerations – are the seeds of the key debates about biopatenting. In many ways, the story of biopatenting is one of the slow expansion of the category of invention at the expense of nature. While biopatenting began slowly and quietly in North America, with patents being issued for a microorganism in the 1960s, it was not until a patent was sought for a life form, as a whole, that biopatenting garnered serious and substantial public, political, and scholarly attention.

The watershed case in North America, which has served as the benchmark for subsequent American and Canadian courts, was *Diamond v. Chakrabarty* (1980). In 1971, microbiologist Ananda Mohan Chakrabarty and his employer, General Electric, applied for a patent on a genetically modified bacterium, which broke down crude oil components. This oil-eating bacterium was potentially very useful (and lucrative) for cleaning up oil spills. Chakrabarty was initially denied by the United States Patent and Trademark Office, and he appealed to the courts. In a close decision, the United States Supreme Court, in 1980, held that the issue was not one of whether the subject of the patent application was animate or inanimate, but rather whether it was made by a person, as opposed to being found in nature. The court found that the bacterium, as a living organism, was a composition of matter, and therefore was a human-made invention. Microorganisms were more akin to chemical compositions than complex organisms. The best-known dictum from the decision, and the one that has had a revolutionary rhetorical and legal impact upon the patenting of life forms, was the favourable quoting of the phrase 'Congress intended statutory subject matter to include anything under the sun that is made by man' (308).

The distinction between nature and culture is left unproblematized by the American high court; yet nature clearly emerges as Heidegger's standing reserve for man as inventor. Nature is rendered invention through human creative agency. Legal scholar Keith Aoki comments on this empowering of the inventor: 'By focusing on human intervention as a crucial factor in determining patentability, the Chakrabarty Court reemphasized the role of the originary inventor who transforms and reshapes the raw materials of the world, thereby justifying a property right in the result. This author-like figure, the inventor, is the subject of patent law' (1993b, 198). However, in addition to refiguring the inventor as subject, and nature as object, *Chakrabarty* is understood by legal, social, and cultural scholars as a significant biopolitical event, a marker of a broader set of shifts that were beginning to crystallize in the late twentieth century. Rabinow remarks, for example, 'The Chakrabarty decision was less a legal milestone than an event which symbolized broader economic, political and cultural changes taking place' (1996b, 132).

Although typically American legal cases are not particularly relevant to Canadian jurisprudence, in the case of patents there are strong similarities between the American and Canadian legislative regimes, including a nearly identical definition of invention. That, combined with the dearth of biopatenting cases in Canada and the powerful symbolic status of *Chakrabarty*, has meant that the decision has garnered substantial consideration in Canada as well. Interestingly, despite the complex ethical, scientific, legal, social, and cultural issues involved in patenting life forms, there have been very few Canadian legal decisions guiding the way. This, along with the deafening legislative silence, has given significant discursive power to the few existing decisions.

In Canada, microorganisms have been property since a Patent Office decision in 1965, but in *Re Application of Abitibi Co.* in 1982, the Patent Appeal Board specifically considered the issue in writing. The board found that a yeast culture that could digest waste product from pulp mills was patentable subject matter. Echoing the *Chakrabarty* court, it held that, because microorganisms are produced en masse in such large numbers, they are analogous to chemical processes. All will possess uniform characteristics and properties and therefore can be perceived as an invention. The board suggested, in passing, that its decision would not likely apply to plants and animals as higher life forms. However, it speculated, 'If an inventor creates a new and unobvious insect which did not exist before (and thus is not a product of nature), and can recreate it uniformly and at will, and it is useful (for example to destroy the spruce bud worm), then it is every bit as much a new tool of man as a microorganism. With still higher life forms it is of course less likely that the inventor will be able to reproduce it at will and consistently, as more complex life forms tend to vary more from individual to individual. But if it eventually becomes possible to achieve such a result, and the other requirements of patentability are met, we do not see why it should be treated differently' (90).

Therefore, in *Abitibi*, invention is determined by analogy from what is clearly scientific, i.e., non-biological. Reproducibility and predictability emerge as central to inventiveness. Through *Abitibi*, residual naturalism becomes an element in the discourses of biopatenting. François Dagognet (1988) suggests that residual naturalism, the ongoing privileging of the natural over the artificial, is due, in part, to how humans have manipulated nature for centuries, although they have not changed it ontologically because the products of human effort do not yet contain an internal principle of generation. In the Patent Board's insect example, we see this link between control over reproduction and artificiality, or invention.

Residual naturalism becomes law when, in 1989, a cross-bred soybean makes it to the Supreme Court of Canada. In *Pioneer Hi-Bred Ltd. v. Commissioner of Patents* (1989), the Supreme Court of Canada was called upon to consider whether or not a complex plant could be an invention under the *Patent Act*. The Federal Court of Appeal had found that complex plants fell outside the language of the *Patent Act*, upholding the Canadian Patent Office's refusal of the patent. The Supreme Court decided the case on other technical grounds, but suggested, in a nonbinding contemplation, that there are two kinds of genetic engineering.

The first is cross-breeding over several generations to produce new varieties. As the court states, 'There is thus human intervention in the reproductive cycle, but intervention which does not alter the actual rules of reproduction, which continues to obey the laws of nature' (6). This is obviously the kind of genetic engineering in which farmers have been engaged for hundreds of years. The second is molecular change, altering the genetic material itself by acting directly upon the gene: 'While the first method implies an evolution based strictly on heredity and Mendelian principles, the second also employs a sharp and permanent alteration of hereditary traits by a change in the quality of the genes' (6). The court concludes that the intervention by Hi-Bred is of the first type, but it was clearly anticipating the cases to come in its discussion of the second.

The court ultimately declined to decide which of the two types of reproduction, if either, is patentable, but offered some significant guidance:

> The intervention made by Hi-Bred does not in any way appear to alter the soybean reproductive process, which occurs in accordance with the laws of nature. Earlier decisions have never allowed such a method to be the basis for a patent. The courts have regarded creations following the laws of nature as being mere discoveries the existence of which man has simply uncovered without thereby being able to claim he has invented them. Hi-Bred is asking this Court to reverse a position long defended in the case law. To do this we would have, *inter alia*, to consider whether there is a conclusive difference as regards patentability between the first and second types of genetic engineering, or whether distinctions should be made based on the first type of engineering, in view of the nature of the intervention. The Court would then have to rule on the patentability of such an invention for the first time. The record contains no scientific testimony dealing with the distinction resulting from use of one engineering method rather than another or the possibility of making distinctions based on one or other method. (6)

Again, the principles of reproduction, and their location within or outside the invention, are determinative. Dagognet comments that this enduring naturalism leads to several cultural axioms, the most relevant here being that the principle of generation furnishes the proof of life, or, as Rabinow puts it, 'life is auto-production' (1996a, 108). This principle of generation – how a being reproduces or is reproduced – becomes a major fault line in the Oncomouse tale.

Pioneer Hi-Bred was the Canadian high court's last opportunity to offer its perspective on biopatenting, until the Oncomouse case was appealed

from the Patent Office, more than a decade later. Throughout the intervening period of the 1980s, the Canadian government was repeatedly
criticized by the fledgling biotechnology industry for its reticence to pursue legislation to bring Canada more in line with *Chakrabarty* and its suc-
cessors. This appeal became louder with the 1988 granting of the
Oncomouse patent in the United States and grew to an outcry when the
Canadian Patent Office repeatedly refused the same patent throughout
the early to mid-1990s.

In 1996, Industry Canada Minister John Manley suggested that, while
the government was studying the issue, it had no plans to bring any policies or legislation forward.[2] Yet, notwithstanding the policy lacunae over
the 1990s, divisible genetic elements of plants, animals, and human beings were repeatedly patented in Canada – cells, genes, gene sequences,
and tissues. Single-celled organisms such as bacteria, some fungi and
algae, cell lines and hybridomas had all received patents. No patent has
yet been sought for the human being as a whole.[3]

In part because there is no law in Canada on the patenting of human
beings, the case of the Oncomouse as the Canadian Supreme Court's first
treatment of the patenting of a 'higher' life form is significant both as a
social science fiction and a biopolitical event. In particular, the case
makes visible truths and authority over the nature, and ownership, of life.
It maps the symbolic terrain of the battle for the objectification of the
human as genetic kin to Oncomouse and implies modes of subjectification particular to the self, as seen through the molecular optic. In addition, the bureaucratic-legal journey of the Oncomouse from the Canadian
Patent Office, to the trial and appeal divisions of the Federal Court, to the
Supreme Court of Canada suggests that the biopolitical implications of
the molecular optic are playing themselves out uniquely in Canada, making Canadian citizens very specific biosubjects.

Setting the Frames: Patent Office

As noted above, the commissioner for patents was the first Canadian
authority to consider the patent application for the Oncomouse, releas-
ing his decision on 4 August 1995. While he allowed the patent on the
process of splicing the oncogene onto the plasmid and injecting it into
the egg, he did not accept the patent on the animal itself. His latter decision turned on whether a higher life form could be considered an invention or, more specifically, a manufacture or composition of matter, within
the *Patent Act*.

While the process of molecular manipulation by the scientists would seem to be much closer to the second kind of genetic engineering, outlined by the court in the *Hi-Bred* case discussed earlier, the commissioner avoided this dilemma by splitting the process into two phases: the gene splicing, and the gestation of the genetically engineered mouse in the uterus of its foster mother. The first phase, he argued, is under the control of the inventor. The second is controlled by the 'laws of nature.' The former is an invention as a manufacture or composition of matter; the latter is not: 'The inventors do not have full control over all the characteristics of the resulting mouse since the intervention of man ensures that reproducibility extends only as far as the cancer forming gene' (7). The power of reproducibility, therefore, continues to be significant. Moreover, the commissioner distinguished cases about bacteria, which dealt with lower life forms, from the Oncomouse, which, as a mammal, is clearly a higher life form. But the commissioner did not elaborate on the distinction between these self-evident and mutually exclusive orders of being.

Three discursive frames are generated by the patent commissioner. First, there is a belief in both a separate domain called 'nature,' and therefore the separability of nature and invention. Second, invention is determined by the degree of control exercised by the inventor over the conditions of reproduction or reproducibility. Third, the decision embraces an obvious and unproblematic distinction between higher and lower life forms. As becomes apparent, these frames significantly structure subsequent debate, because, within the legal system, the previous level of decision is always the discursive frame within which the next authority begins. Further, the talk remains solely bureaucratic and legal, as the Patent Office's decision remains under the radar of the Canadian media.

Authorizing the Frames: Federal Court, Trial Division

Harvard appealed the decision of the patent commissioner to the Federal Court. At the Trial Division, in a decision released on 21 April 1998, the court agreed that the Oncomouse could not be patented (*Harvard College v. Canada* 1998). It acknowledged that the rodent was new, useful, and unobvious – all points that were not in dispute between the parties. However, that did not resolve the primary question: is the mouse an invention? Justice Nadon defined four key issues that went to the mouse's status as invention. First, 'Is it appropriate to examine the degree of the inventor's control over the creation of the claimed invention?' (para. 20); second, 'Is it appropriate to distinguish between human intervention and the laws of nature?' (para. 25); third, 'What is the relevance of the test of reproducibility in the present instance?' (para. 31); and fourth, 'Is it appropriate in determining whether something is patentable subject-matter to make distinctions between higher and lower life forms?' (para. 33).

On the issue of the degree of control of the inventor, the court found that it was not necessary for an inventor to control all the natural processes leading to the end product. Nevertheless, the mouse in this case was found to be completely unknowable and unknown: 'There is no way to separate the transgene from the rest of the mouse once it is introduced and everything else about the mouse is present completely independently of human intervention' (para. 24). On the second question of the relationship between the inventor and the laws of nature, the court found, 'The creation of the oncomouse is a marriage between nature and human intervention' (para. 27). Because the outcome is infinitely variable and unknown, it is more nature than invention.

On the third question, the court held that the mouse is not reproducible under the *Patent Act* because 'too much is left to luck and chance' (para. 32). The breeding of the mouse takes place in the ordinary manner, and the inventors are not producing it at will. The court suggested, 'Although the gene will be present in some mice, at some place, with some characteristics, the precise mouse, the precise location and the precise quality of the gene are irreproducible. The variations of the gene are created and controlled completely by the laws of nature and are infinite' (para. 32). Finally, on the question of the appropriateness of distinguishing between higher and lower life forms, the court concluded that it is appropriate to do so; that Parliament should decide the issue; and that the court itself would find higher life forms not patentable.

As a result, the discursive frames established in the Patent Office continue. First, nature and culture are different zones that can be distinguished, and appropriately so. Second, the inventor's ability to control the unpredictable laws of natural reproduction determines invention. Third, a self-evident distinction between higher and lower life forms is possible and appropriate. But, a fourth frame is added: the mouse, as a complex being, is found to be indivisible, in that the transgene cannot be separated from the rest of the mouse. And this time, the discursive frames have judicial authority.

Interestingly, given the significance of the issues addressed by the court, the press coverage of the Federal Court decision was, again, negligible.[4] While it is impossible to draw any strong conclusions based on such a small sample of media, a few themes are visible, which are repeated in subsequent press coverage. In the media, ethical debate enters into the discussion much more openly than it can do in the legal decisions, where it is virtually absent. There is also an even less veiled critique of the federal government's lack of legislative or policy action on the issue, than is found in the court's discourse (e.g., Walkom 1997). Ultimately, there is a real sense in the coverage that this is the state of the law, *for now*. The seeming inevitability of both genetic science's progress and legal appeals is reflected within the media treatment. All social actors appeared content to wait for the next battle in the judicial war, one that took place in 2000.

Renegotiating the Frames: Federal Court of Appeal

Harvard appealed the decision to the Federal Court of Appeal, where they finally won on 3 April 2000 (*Harvard College v. Canada* 2000). The majority decision found that the Oncomouse was a new, useful, and unobvious 'composition of matter,' and, therefore, an invention under the *Patent Act*. The court concluded that if the artificial oncogene sequence is a composition of matter (which was not in dispute), then all offspring with that gene would also be compositions of matter, regardless of how they were (re)produced. In short, the Oncomouse is not found in nature. The court stated, 'The question then is whether the Oncomouse is merely a discovery of a natural phenomenon or involves the application of inventive ingenuity' (para. 47). In a break with the previous discursive frames, here, while control is an element of inventiveness, complete control over reproducibility is no longer required: 'Usefulness is necessary for patentability and implies control in the sense that the desired result will be achieved when the product is used or produced. That desired result in this case is an oncomouse with susceptibility to cancer for use in carcinogenicity studies. Once that has been achieved, control over other characteristics of the mouse is not necessary or relevant. Such "additional" control has nothing to do with the desired result' (para. 78). In this, the Federal Court of Appeal broke with the naturalist frame, suggesting that any tampering with auto-generation is adequate to de-naturalize both the process and the resultant organism.

Further, the court declined to divide the process and the product, as all previous courts and administrative bodies had done. 'If the process for producing the product is patentable, it is because it must be considered to involve ingenuity ... It must logically follow that the product of that process must also be considered to involve the same ingenuity and be patentable' (para. 48). The majority decision rests on the assumption that the organism, itself, is divisible, that it is only the sum of its parts. It found that the Oncomouse is a composition of matter because the transgenic unicellular material that was transferred into the host mouse was a composition of matter. The court disallowed any kind of transformative change in the natural reproductive process. Reproductive change, in this decision, is a quantitative, not a qualitative, development, disciplined by the application of human ingenuity.

The interpretation of the mouse as divisible permits the appeal court to reject the easy separation of nature and culture. Because the offspring contain a trait that is not found in nature, a trait that is based on a composition of matter, the mouse itself is a composition of matter, regardless of whether it receives its oncogene through genetic inheritance, or through human intervention (para. 42). In other words, an element of human invention transforms the natural into the cultural, rendering a higher life form an invention, an effect of the operation of biopower.

Lastly, the court made a distinction between policy and law, noting, 'There was considerable fanfare in this appeal that significant policy questions are at stake ... To the extent that the appeal gives rise to policy questions, they are to be addressed by Parliament and not the Court' (para. 30). The judge went on to say, 'In this type of case, the Court is not the forum for a public policy debate' (para. 117). While it might not be the ideal forum, in a typically Canadian twist, the courts have increasingly become the forum in which the terms of the debate are framed, and it is in response to those frames that the mediated public debate is mobilized.

Because this decision changed the status quo – suddenly higher life forms were patentable in Canada for the first time – the press coverage of the Federal Court of Appeal decision was more substantial in amount, if not in substance. The result was heralded repeatedly as a 'major victory' for Harvard and for the biotechnology industry. Canada was seen as finally coming into line with Europe and the United States in economic competitiveness and patent regulation. The biotechnology industry, represented both by several high-profile patent lawyers and by BIOTECanada (the national umbrella organization of most Canadian biotechnology companies), received considerable attention. Critics of the decision were harder for the media to locate, given the nature of this particular dispute. Typically, a legal conflict would pit advocates of two opposed positions; in this case, Harvard's opponent is CIPO, which is defending an administrative appeal of one of its decisions. As a result, the dramatic potential of this case, filled as it is with the potential for heroes, villains, monsters, mad scientists, and slave species, is barely registered. Instead, it is presented as a rather mundane story about patent holders where venture science appears as the only real social actor worth considering.

What is perhaps most striking about the press coverage is the absence of widespread debate on the larger ethical and social issues. While a *Calgary Herald* journalist predicted, 'The landmark decision from the Federal Court of Appeal is expected to spark a fierce debate over bioethics and whether humans should have the right to patent other living things' (Tibbetts 2000), this debate simply did not happen. It is this absence that is most compelling and raises important questions about biosubjectivity and biopower in the Canadian context.

At the time of the ruling, *Canadian Business* contended, 'There are no easy answers. But apart from isolated discussion in pockets of the academic, scientific and legal communities, there's been little public debate on human gene patenting in Canada' (McClearn 2000, 129). The call was issued: 'Canada needs a full Parliamentary review of whether to grant patents on higher life forms and, if so, with what conditions and safeguards' (Christie 2000). Two back-to-back editorials in the *Ottawa Citizen* sought a parliamentary response. As the 11 August editorial stated, 'If biotechnology is scary, let Parliament save us from Frankenfoods and killer tomatoes' (2000).

As was also the case with reproductive technology, the government was strongly criticized for continuing to avoid the thorny issues of life, health, bodily integrity, and biopower that biotechnology had wrought, repeatedly described as hiding behind judges' robes: 'We elect our representatives to publicly debate issues of national importance and it seems clear that the implications of patenting life forms and the impact on medical research are very real and pressing concerns affecting all Canadians. But these and other publicly relevant questions were relegated to the courts because they were deemed too controversial, too divisive and, dare we say, too mentally challenging for the politicians, who prefer to bore us with unity debates and internal squabbles over who's giving up his or her seat for the next by-election' (Boyd 2000). Editorials such as these were particularly biting in their critique of the government. The editors of the *Calgary Herald* suggest, 'If Parliament wants to prevent patenting complex life forms it must change the law, not expect the courts to do its dirty work ... Genetic engineering is here to stay. If Parliament wants to keep Canada out of the research loop, then it should bear the full brunt of that decision. It should not make the courts a convenient scapegoat to sidestep the debate' (2000).

Maureen McTeer, having emerged as an activist on questions of reproductive technology, offers the highest profile call for public debate by suggesting that 'the Federal Court's decision in the Harvard Mouse patent case will give us all a lot to talk about. This discussion is finally back where it should be, in the public domain, where Canadians can decide how and what they want to encourage and prevent in the exciting new world of genetic engineering of life forms' (McTeer 2000).

Yet whereas the abortion issue or reproductive biotechnology received some attention – often thwarted or deliberately misdirected – the discussion never even began. The acting patent commissioner of the day repeatedly indicated his satisfaction with the existing legislation. Industry Canada Minister John Manley was also constantly quoted, indicating that his office was considering no changes to the *Patent Act* in the patenting of higher life forms. At the time of the Federal Court of Appeal's decision, approximately 250 patent applications for animal life forms and another estimated 350 for plants were on hold at CIPO, pending the outcome of the decision.

Clearly, the decision of the Federal Court of Appeal rejects the residual naturalism of previous decisions, challenging the separability of nature and culture, affirming the divisibility of the organism, and eliding the distinction between higher and lower life forms. However, given the likelihood of an appeal to the Supreme Court of Canada, this renegotiation of the frames of Canadian biopatenting was contingent and uncertain. More significantly, the court marked out a domain of truth about which it is competent to speak – the technical interpretation of patents, and the relationship between nature and culture – all the while denying its own competence to speak to other issues: ethical, political, and social. This, of course, implies that issues such as the separation of nature and culture, the determination of the divisibility of an organism, and so on, can be separated from their ethical, political, and social implications. Such a separation of law and politics is further endorsed by most pundits and journalists weighing in on the Federal Court of Appeal's decision. The authority of the courts to speak on the social, political, and ethical issues was challenged, and a call was issued to another social institution to speak that truth – Parliament.

Rewriting the Frames: Canadian Biotechnology Advisory Committee

In the wake of the Federal Court of Appeal decision, a letter to the editor of the *Ottawa Citizen* from the National Council of Women of Canada stated, 'We believe the government should set up, as soon as possible, a task force or parliamentary committee to consider these questions, ensuring that the results of basic research in this field remain in the public domain' (Hutchinson 2000). Ironically, such a body already existed at the time, but clearly it had failed to make its profile well known to Canadians and it had not served as the nodal point for any kind of public debate.

The Canadian Biotechnology Advisory Committee (CBAC) was the federal government's attempt to address the calls for public debate and for parliamentary action on biotechnology. CBAC was created in September 1999 as part of the government's broader Canadian Biotechnology Strategy to provide guidance on policy issues in the field. It was composed of about twenty external experts drawn from law, the biotechnology private sector, nutrition, agriculture, medicine, molecular biology, and so on. It reported to the Biotechnology Ministerial Co-ordinating Committee, which comprises the federal ministers of industry, agriculture, and agri-food, health, environment, fisheries and oceans, natural resources, and international trade.

In early 2000, CBAC initiated research and consultation to investigate the implications of patenting higher life forms. Its final report, entitled *Patenting of Higher Life Forms*, was released in June 2002. The report acknowledged up front that the Oncomouse case, which was before the courts, generated the consultation process (CBAC 2002a, 1). While these issues surrounding the patenting of higher life forms were seen to be generally significant, the Oncomouse, as a biopolitical event, served as a key catalyst in the administrative response. As one journalist aptly phrased the issue, 'Build a better mousetrap and the world will beat a path to your door. Build a better mouse and the bureaucrats will come knocking' (Campbell 1996).

CBAC defined the key issues as follows:

- Ought higher life forms be subject to patent rights?
- If so, what measures are needed to protect the dignity of and maintain respect for human beings?
- If patent rights are extended to plants and animals, what ought to be the scope of those rights, taking into account their particular nature?
- How can the patent system be made more effective with respect to higher life forms?
- Does the intersection of biological inventions and patent law raise other issues that need to be addressed, whether in the patent system or elsewhere? (2002a, 3)

Hence, the CBAC process appeared to promise that the robust public debate sought by courts, journalists, and analysts would be forthcoming, finally.

The committee commissioned a number of research reports, reviewed public opinion research, and organized a series of three 'stakeholder meetings,' with stakeholders defined as non-governmental organizations, scientists, and industry.[5] On the basis of these meetings, CBAC prepared a consultation document, entitled *Biotechnological Intellectual Property and the Patenting of Higher Life Forms: Consultation Document 2001*, in April. It then conducted limited and targeted consultation with stakeholders, receiving virtually no public feedback. On this basis, in November 2001 CBAC prepared the *Interim Report on Biotechnology and Intellectual Property*, in which it announced its primary conclusions. Individuals and organizations had until 15 March 2002 to offer their responses. The final report was released in June 2002, not, incidentally, in time to be within the purview of documents reviewed by the Supreme Court of Canada as it heard the Oncomouse appeal.

Yet, after this long, administrative consultation, in the summary of responses to the interim report, CBAC was repeatedly careful to note that the respondents, who participated in the formal consultations and responded to the report, were primarily those with a direct stake in biotechnology: 'They were members of the biotechnology industry, users of these technologies, industry or university researchers, or funders of research in the biotech field. The reader should bear in mind that this group of commentators is not a random sample of the Canadian public. The strength of support for any particular view described in this document does not, therefore, necessarily represent that of the Canadian population' (2002b, 1). Thus, by CBAC's own admission, they had failed to engage the general public in Canada in any meaningful way. This limited engagement, itself, generated media comment. Colin Freeze of the *Globe and Mail* commented wryly on CBAC's online questionnaire, which posed several questions: Should mice be patented? How about higher life forms? Which ones exactly? Where should the line be drawn? He noted, 'A new questionnaire asks the Canadian public to ponder these head scratchers ... It's a bizarre exercise in public opinion gathering. But this is, in essence, what the Canadian government is doing through an arms-length biotech advisory agency' (Freeze 2001).

As is the case with so many Canadian government consultations, the format, structure, and nature of the process results in participation by direct stakeholders and excludes the population at large. As a result, conclusions and recommendations are predetermined and robust exchange over ethical and social concerns is neatly avoided. The consultation did not seem to register any broader public awareness, as can be seen in the limited press coverage. The final report does not fare much better. As a *Canadian Business* reporter observes, 'Earlier this summer, the Canadian Biotechnology Advisory Committee (CBAC) submitted a report that's crucial to the future of the biotech industry in Canada. It was received with very little fanfare and, except for the odd newspaper story, the CBAC's final report on patenting higher life-forms went pretty much unnoticed. That's surprising. Given the uproar over cloning, stemcell research and genetically modified foods, you'd think a long line of politicians and commentators – reasonable and otherwise – would be ready to weigh in on the controversial subject' (Wahl 2002, 75).

Again, however, one of the most significant points to emerge from the coverage was the widespread support of the committee's call for broader public debate on biopatenting, but with no corresponding agenda or system to see that it took place. The CBAC process, while interesting and in-depth, did not offer the widespread debate that pundits, journalists, and critics were requesting when the Federal Court of Appeal decision was released. Rather, it served as a forum for institutionally endorsed experts and interested parties to speak in a discursive site always already privileged to influence state (in)action. In other words, it provides an exemplar of the process of normalization within biogovernance that we discuss in chapter 1.

CBAC played a privileged role in the discursive framing of the biopolitical event of Oncomouse. It addressed and yet contained the ethical and social issues of biopatenting, framing them as external to the knowledge system of patent law, confined to an appendix in the Report. Patent law is framed as a technical way of knowing. As such, the final report merely reaffirmed Parliament's role, calling for fast action and challenging the legal system's authority to regulate life. In what by now had become a monotonous and hollow call for accountability, the report states, 'Given the importance of these issues to Canadian society generally and to health care and agriculture in particular, as well as the significant "values" content of the issues raised, we believe that Parliament and not the courts should determine whether and to what degree patent rights ought to extend to plants and animals' (CBAC 2002a, 7). Despite their own call for more public input, CBAC's own report offered only scientific and industrial expert endorsement of the decision that Oncomouse could/ should be patented. Further, given that CBAC is a non-legal institution, it could frame the 'pro-patenting' position in a richer language than the courts. As a result, it recommended that human bodies not be patentable in any stage of development, but that all non-human life forms be patentable, subject to certain provisions to protect special interests, including farmers, innocent bystanders, and researchers. Higher life forms were defined by CBAC as plants, seeds, and non-human life forms, other than single-celled organisms (CBAC 2002a, 6).

The committee attempted to resolve the issue of reproducibility in the same terms as the Federal Court of Appeal. Once a life form has been touched by an inventor, it is transformed into an invention, regardless of the mode of reproduction: 'Because higher life forms can reproduce by themselves, the grant of a patent over a plant, seed or non-human animal covers not only the particular plant, seed or animal sold, but also all its progeny containing the patented invention for all generations until the expiry of the patent term' (CBAC 2002a, 12).

Of course, the big issue on everyone's mind was not whether mice could be patented, but whether humans could be. The courts had painstakingly avoided speculation on this minefield, but CBAC did not. In doing so, the committee was very precise in its language. The interim report had used the terminology *human being* in relation to patenting, whereas the final report carefully substituted *human bodies*. This shift was because '[a] human being is a metaphysical concept, not a biological one. The substitution of the word "body" for "being" eliminates this awkwardness' (CBAC 2002a, 9). The plural of *human bodies* was also used very deliberately: 'By using the plural, emphasis is placed on the whole human body and not on its parts (for e.g. artificially created human organs). Thus, the phrase "human bodies at any stage of development" is more likely to be read narrowly – as we intend. It is important not to discourage research on stem cells and the creation of artificial organs' (CBAC 2002a, 9). The use of *human bodies*, therefore, implies divisibility and the end of human bodily integrity. On a metaphysical level, then, CBAC ruled that humanness is not vested in the body.

Despite their attempt to sidestep a complex problem of patenting higher life forms, the legal basis for the non-patentability of the human being or an entire human body remains unclear. In a footnote, the committee notes, 'Even though human beings are animals, most lawyers maintain that a whole human being is not patentable, or else that patents over whole humans would not be enforceable' (CBAC 2002a, 6). Subsequently, the committee adds, 'It is generally believed unlikely that a holder of a patent over a human DNA sequence or cells (including stem cells) would be able to exercise control over a human body containing that sequence or cell. Nevertheless, the law has never explicitly addressed the issue' (CBAC 2002a, 8). The basis for this surmise is the principle of respect for human dignity. 'Even if the act of granting a patent on an invented human were not in itself a violation of basic human rights, exercising the patent's exclusive right to make, use or sell an invented human *would almost certainly* violate the *Canadian Charter of Rights and Freedoms* and the *Canadian Human Rights Act'* (emphasis added; CBAC 2002a, 8).

These claims are important for a variety of reasons. They seek to locate the prohibition against the patenting of a human being in positive, rather than natural, or common law; they acknowledge the possibility of an 'invented human'; they grant that humans are animals. They frame the problem more around the patenting of the human than his/her/its invention. They highlight the unstable basis of the legal authority against patenting human beings in Canada. And they locate the primary objection in human dignity, a basis whose efficacy has been challenged by a number of scholars (e.g., Mitchell 2004; Pottage 1998). Finally, they attempt to smooth over any troubled waters with the rather weak assumption that the courts would 'almost certainly' rule against any attempt to own a human body or being.

Thus, in its report, CBAC endorsed and supplemented the discursive frames renegotiated by the Federal Court of Appeal within the legal system when it ruled in favour of Harvard. The truths produced in CBAC's language were not only richer and more detailed, they deployed and gained legitimacy from a variety of bases of expertise: legal, scientific, industrial, and social. Moreover, CBAC raised the spectre of patenting humans just in time for the final appeal to the Supreme Court. While the court did not tackle that issue directly, its decision now necessarily had implications for all biosubjects, be they mice or men.

Supreme Court of Canada: Defining Biopolitical Truths

The 5–4 decision of Canada's highest court in favour of the commissioner for patents was released on 5 December 2002 (*Harvard College v. Canada* 2002). In the preceding months, however, CBAC was busily completing its work, and other concerned parties were also active: Industry Canada, for example, commissioned a focus group study that revealed anxiety among a majority of Canadians about the patenting of genes, organs, and higher life forms. They trusted neither Parliament nor the courts to deal with the issue. Indeed, only after extensive 'education' by the researchers did some of their concerns abate. This in itself raises questions about how public opinion is not mobilized in order to generate debate, but is managed to quell the possibility of that debate.

BIOTECanada was also busy, attempting to influence the discursive construction of the issue. In a press release issued prior to the Supreme Court of Canada decision, it tried to shift the frames of debate to those of industrial viability, global competitiveness, and health. BIOTECanada stated, 'With the patent currently under appeal, Canadian researchers are left in doubt about the patentability of other higher life form innovations. This discourages innovation of potentially life-saving discoveries' (in Canada NewsWire 2001). The BIOTECanada statement posed a marked contrast to a poll conducted of biotech researchers. In November 2000, Environics Research Group surveyed the Canadian biotech community and released its results in March 2001, revealing an abysmal level of knowledge about patents among Canadian researchers. A majority were confused over what constituted a research exemption. Twice as many members had a low understanding, as had a high comprehension, of patent law. The study also suggested that concerns about patent infringements were inhibiting research at the early stages. One-third of respondents reported that they had delayed their research work; another third reported postponing work; and one-fifth said they stopped work completely, supporting the critics of patents, who claim that patents stifle rather than enhance the research environment. Only a third of Canadian biotechnology researchers surveyed knew the outcome of the Federal Court of Appeal decision in the Oncomouse case.

Finally, in response to the CBAC final report, it appeared that Industry Canada might actually take action prior to the Supreme Court of Canada decision. In October 2002, a government official indicated that the federal government would be clarifying its position on the patenting of higher life forms and genes, prior to the Supreme Court of Canada decision. The government even promised to revamp the *Patent Act* in its October 2002 Throne Speech. Ultimately, no action was taken prior to the decision. And, despite the combined efforts of CBAC, BIOTECanada, and Industry Canada, when faced with the issue, the Supreme Court allowed the appeal, stating that a higher life form was not patentable because, on a plain reading of the words in the *Patent Act*, it was not a 'manufacture' or 'composition of matter.' This was the ruling for which everyone had been waiting since the Oncomouse patent application had been filed seventeen years before.

The majority of the court held that, upon a straightforward interpretation of the legislation, Parliament did not intend higher life forms to be patentable: 'Given the unique concerns associated with the grant of a monopoly right over higher life forms, it is my view that Parliament would not likely choose the *Patent Act* as it currently exists as the appropriate vehicle to protect the rights of inventors of this type of subject matter' (para. 120). The majority agreed that, while the fertilized egg could be construed as a composition of matter, the mouse as a whole could not: 'The body of a mouse is composed of various ingredients and substances, but it does not consist of ingredients or substances that have been combined or mixed together by a person. Thus, I am not satisfied that the phrase "composition of matter" includes a higher life form whose genetic code has been altered in this manner' (para. 162). The majority asserted that higher life forms have qualities and characteristics that transcend their genetic composition.

The Supreme Court endorsed making the distinction between higher and lower life forms but failed to provide any definitions. As a result, one is either a higher or lower life form with the resulting struggle to inhabit one or another of those mutually exclusive and apparently self-evident categories. Very different consequences result from one's categorical presence – lower life forms can be rendered property, higher life forms cannot.

The minority at the Supreme Court opted to distinguish humans from animals, asserting, 'There is a qualitative divide between rodents and human beings' (para. 102). But it refused to endorse a distinction between higher and lower life forms, suggesting that those boundaries are policy driven and, therefore, not for the court to draw. The majority went further, maintaining the broad categories of higher and lower life forms. They suggested that the distinction is one anchored in common sense differences between the two, adding that if the line between higher and lower life forms is indefensible and arbitrary, then so, too, is that between human beings and other animals. The analogy between lower life forms and compositions of matter was made easier by the majority's generalized definition of lower life forms that included their capacity to be produced en masse as well as their uniform properties and characteristics. But the Supreme Court again went further than the other social actors: 'In particular, the capacity to display emotion and complexity of reaction and to direct behaviour in a manner that is not predictable as stimulus and response is unique to animal forms of life' (para. 204).

Ultimately the high court's distinction is a resort to the clean lines of demarcation that mutually exclusive categories create, whether they be higher and lower life forms, human and animal, or nature and artifact.

Yet, as Haraway suggests, 'It will not help – emotionally, intellectually, morally or politically – to appeal to nature and the pure' (1997, 62). Biotechnological practices are complicating the simple categories that are preserved in the majority decision. Ironically, the court itself recognized this point when it speculated on xenotransplantation. 'The pig receives human genes. The human receives pig organs. Where does the pig end and the human begin? How much DNA does it take before one becomes the other? The answer to these questions once ridiculous and offensive, may now just be a matter of degree' (para. 180). Nonetheless, in its refusal to collapse the categories of higher and lower life forms into human and non-human (or completely), the Supreme Court renders humans kin to other mammals and multi-cellular, complex organisms. Singular human status is rejected for a model that does not deny the animal in the human. Humans and mice are equally biosubjects.

The dissenting judges disagreed, favouring a more technical approach and allowing economic, but not moral or political, issues to enter. Justice Binnie contended, 'The legal issue is a narrow one and does not provide a proper platform on which to engage in a debate over animal rights, religion, or the arrogance of the human race' (para. 1). The technical simplicity is anchored to the definition of invention: 'I do not think that a court is a forum that can properly determine the mystery of mouse life. What we know, in this case, is that the inventors were able to modify a particular gene in the oncomouse genome, and produce a new, useful and unobvious result. That is all we know about the mysteries of oncomouse life and, in my view, it is all we need to know for the purposes of this appeal' (para. 78). However, as Haraway aptly notes, 'The messy political does not go away because we think we are clearly in the zone of the technical, or vice versa' (1997, 68).

After the decision was released, the Patent Office rejected many of the 1,500 applications that had been on hold pending the Oncomouse decision. These included a salmon genetically altered to grow quickly; a goat that produced milk with drug proteins in it; fish engineered to secrete human insulin; a cow designed to produce human milk; a non-allergenic peanut; a disease- and pest-resistant poinsettia; as well as forgetful and blind mice being developed to study Alzheimer's disease and macular degeneration, respectively.

Intervening groups were very pleased with the outcome, but the biotech industry was not. One journalist described the industry as 'decidedly furious' (Starr 2003, 38), and the decision was called a 'body blow' to biotechnology (Klein 2002). Spokespeople for biotech industry organizations across the country posed their disagreement in threatening and fatalistic terms. The president of AgWest Biotech, Peter McCann, said, 'This is a very bad day for Canada, quite frankly' (Klein 2002). An official from the University of Saskatchewan, responsible for commercializing research, warned, 'This sets a reactionary tone for Canada' (Klein 2002). Janet Lambert, president of BIOTECanada was most vocal: 'This decision stops our pursuit of knowledge and innovation dead in our tracks ... Today's decision destroys our Canadian infrastructure of knowledge and innovation, creates an even greater brain drain, and we will lose our place at the world table in influencing how and where society accepts this technology' (in Canada NewsWire 2002). Editorialists in conservative papers picked up the frame of global competitiveness: 'The decision sends an unfortunate and probably unintended, message that Canada is afraid of biotechnology. If this nation wants to thrive in a globalized, research-based economy, it must be open for business, and spurning the mouse patent suggests we aren't. The international scientific and business communities have followed the fate of the Harvard Mouse closely, and may now consider Canadians the Luddites of the industrialized world' (*Ottawa Citizen* 2002).

In addition to the struggle to frame public discussion of the issue solely in terms of economics, the decision and the ensuing press coverage once again highlighted the question of judicial authority, and the absent voice of Parliament. CBAC had introduced its report with the claim, 'Once the decision of the Supreme Court of Canada in the Harvard mouse case is known, no matter what the ruling, the federal government will have its own decisions to make' (2002a, ix). Repeatedly, the federal government was admonished and exhorted to speak. The Supreme Court majority wrote, 'Whether higher life forms such as the oncomouse ought to be patentable is a matter for Parliament to determine. This Court's view as to the utility or propriety of patenting non-human higher life forms such as the oncomouse are wholly irrelevant' (para. 153). Both majority and minority seconded CBAC's call for Parliament to resolve the question.

This call was echoed vociferously in the press. Just as was the case with reproductive technologies, discussed in the previous chapter, it appeared that Parliament was deliberately avoiding a public debate by offloading the issue onto the courts as a way of limiting the debate to be strictly legal – not ethical, social, or political. One journalist recognized the lack of ongoing public debate: 'More important [than the outcome at the Supreme Court itself], why haven't we been talking about this more openly for the past 17 years? And why now that we are thinking about it, is the issue being debated by nine justices instead of by 301 parliamentarians?' (Hrabluk 2002; see also *Vancouver Sun* 2002).

The imperative for Parliament to act and act quickly was reiterated repeatedly: 'Parliament cannot avoid the question of what constitutes a higher life-form now that the Supreme Court has ruled a genetically modified rodent can't be patented' (Gervais 2002). An editorial in the *Ottawa Citizen* suggested, 'The court has effectively sent the problem to Parliament. This is where it belongs, so let the debate rage, quickly. Biotechnology is an industry of rapid growth; if we want to play, we need sensible, 21st-century rules ... So modernize Canada's patent legislation, quickly. Don't let this tale end badly' (*Ottawa Citizen* 2002). Overall, the press recognized the larger issues at stake: 'There must be a middle ground between relegating Canada to a scientific backwater and allowing a bio-engineering free-for-all to create a latter-day version of *The Island of Dr Moreau*. Surely, what is or is not eligible for the economic protection of a patent is something open to agreement by people of common sense and good faith. And isn't that what our legislators are supposed to be? It's time for Parliament – not the courts – to make some tough decisions, and some law. It's time to set the rules for the bio-engineering age' (Chidley 2002, 8).

In this climate, the government could hardly fail to respond. In questioning in the House of Commons, Industry Canada Minister Alan Rock said he would act 'soon': 'We have to decide how to balance, on the one hand, how to encourage research and innovation and, on the other, to reflect the values of Canadians. Whether you can patent a higher life form raises issues about how we regard life. We have to come to grips with those issues. We can't be afraid of them' (in Kirkey 2002). Despite the pledge for action made in response to the Supreme Court's decision, a year later the government still had taken no action on the issue of biopatenting higher life forms. New legislation was, of course, not tabled before either the 2004 or 2006 elections, and there remains, to this day, no hint that the government will move ahead on this agenda soon.

This lack of political action produces a gap in governance, which is, of necessity, being filled by the courts. The courts are in the position of both deciding the policy issues, despite their protestations to the contrary, and yet undermining their own authority to do so, in those very judgments that appeal to Parliament's superior authority on policy matters. This phenomenon is not particularly unique to the biotechnology domain but is part of a wider governmental shift in Canada. Yet, in the case of Oncomouse, it has specific implications. The law remains a privileged discourse of prescription in society, marking norms of behaviour. However, in this case, it is vitality and identity, rather than behaviour, that are being prescribed, and by the authority of the highest court of the land, a voice that remains authoritative in the absence of government speech.

Consequently, what began as mere discursive frames in the Patent Office and the lower courts – the distinction between higher and lower life forms, control over reproducibility, the separation of nature and culture, and the divisibility of an organism – construct a series of biopolitical truths through the institutional authority of the Supreme Court: truths about the nature of life, identity, and property. These truths figure the central debate within the case, the contestation around subjectivity: the contest between genetic versus organic selfhood.

The genetic model of selfhood – advocated by the Federal Court of Appeal majority, CBAC, and the Supreme Court minority – endorses the potential to reinvent the non-human higher life form. It assumes the divisibility of the body, the irrelevance of any non-human identity, and the capacity for the 'sum of the parts' to be added up by human, rather than only natural, agency. Yet this model does not go unchallenged. The majority decision at the Supreme Court claimed that higher life forms are not fully representable by their genetic profile or elements. 'It also is significant that the word "matter" captures but one aspect of a higher life form … Higher life forms are generally regarded as possessing qualities and characteristics that transcend the particular genetic material of which they are composed' (para. 163). The transcendence of genetic practices and even identity is located in the sanctity of the body. For the majority of the Supreme Court, Oncomouse's body serves as a barrier to the divisibility of her/his/its identity. Both self and body are more than the sum of the genetic parts that they comprise. Hence, a modernist conception of identity is intimately connected to the body, which is conceived as a whole. And yet, as discussed in previous chapters, the body as an entity, concept, and bulwark has become unstable and untenable – in other words, contingent.

As a result, biopatenting emerges as a highly productive site of biosubjectivity. This continues in the *Monsanto Canada Inc. v. Schmeiser* (2004) decision discussed in the Conclusion, which, although partially undoing the legal outcome of Oncomouse, does not change the terms of reference of genetic versus organic selfhood, of biosubjectivity. Yet biopatenting – combining, as it does, both science and property law – produces a particular type of biosubject, a biojuridical subject, which can be understood, as can be the self, by implication, as property.

Biojuridical Subjectivity

The very viability of biopatenting as a scientific-legal technique turns on its ability to render biological material and beings property. Yet property, despite its colloquial understanding as something tangible and knowable, is a very unstable concept. This is particularly the case with intellectual property. As Aoki notes, the 'Rorschach-like tendency of property to radically change shape depending on one's perspective is further exacerbated in the realm of intellectual property' (1993a, 24). Rather than thinking of property as a state of being, or a final outcome or product, the mutability of property requires that it be considered as a process. The subject/object of biopatenting is, therefore, one amenable to propertization, to being conceived as a variously configured bundle, in this case, of intellectual property rights – rights that accrue value, are alienable, and, above all, require management.[6]

But propertization of the higher life form is no easy task. Anti-slavery laws, anti-cruelty laws for animals, human rights declarations, religious doctrine, legal notions of bodily integrity and dignity of the person, and so on, all contribute to a matrix of discourse that marks certain lives not amenable to characterization as property. As Pottage argues, 'Most critiques of biotechnology patents seek to restore the proper bounds of property; the common basic practical response consists in the "juridification" of human life, granting it a form of legal immunity from commodification' (1998, 153). This often results, as is apparent in the Supreme Court of Canada decision, in the reassertion of nature as 'the correlate of sovereign legal power' (155). And yet, in biopatenting, those distinctions come under scrutiny. No longer taken for granted, the exact boundaries of property and life, between persons and things, are rewritten in the work of biopatenting processes.

Repeatedly, beings and their elements, not metaphysically conceivable as things, are rendered such through the pixie dust of patent law. How did this come about? How is it that CBAC can speak of the 'invented human' in a manner that is neither a priori nonsensical, nor abhorrent? Other examples already exist in both medical and cultural frameworks. For example, a cell line from a patient's cancerous spleen was declared 'owned' by the doctor who developed it, but not by the patient, John Moore, himself (*Moore v. The Regents of the University of California* et al. 1990). The quagmire of uncertainty that biopatenting creates is given the absurd treatment it may well deserve by British poet Donna Maclean's provocative act to file an application for a patent on herself, citing her self as original, the product of much inventive effort, and indeed very useful. The framework of responsibilization comes into view when we begin to think of bodies as component parts available for reinvention as property.

It is not self-evident how the higher life form, and its resulting elevation to property, shifts the philosophical grounding of modernity to a materialized reality requiring management (see Davies and Naffine 2001). Critics who lament life forms as property or commodities begin with the outcome, failing to recognize the integral conceptual shifts that precede such a state of thought. In order to read the higher life form as thing, subject to interpellation as property, a number of conceptual shifts first must have taken place, we argue – shifts accomplished at the curious but powerful juncture of the courts and science. These shifts begin to map new ways of thinking about the self. We name these: division, isolation, representation, authorization, and enforcement.

Division

The first conceptual shift required to accept the higher life form as property is that the subject/object of property must be able to be divided into its components parts. Such a division solves the problem of the untouchable, ontologically protected 'whole.' Therefore, the organism is composed of cells, DNA sequences, proteins, and so on, which can be treated separately from the organism itself. Finn Bowring (2003) contends that this is the 'dominant fiction' underlying and legitimizing genetic engineering. It is the next logical step in the contingent body: the contingent self. Rabinow points to the consequences of such a genetic understanding for the person: 'The approach to "the body" found in contemporary biotechnology and genetics fragments it into a potentially discrete, knowable, and exploitable reservoir of molecular and biochemical products and events. By reason of its commitment to fragmentation, there is literally no conception of the person as a whole underlying these particular technological practices' (Rabinow 1996b, 149). By subjecting the body to the molecular optic, the body is effaced as an indivisible entity, and so is the person along with it. Each of the fragments produced in this process can then be treated, removed, isolated, altered, and, of course, patented. Their cumulative effect can be ignored. While most commentators hasten to add that outside of history and science fiction no one can own another person, the same claim is not made about the genetic bits of the person. In order to understand the person as property, the person must be divisible into less metaphysically powerful elements. The synecdochal effect of the person must be disrupted.

Isolation

Because one cannot patent something that is found in nature, how is it that genes, as they are found in the higher life form's body, are patentable? Richard Gold articulates the scientifically endorsed legal fiction (or is it a legally endorsed science fiction?) – isolation – at the heart of biopatenting: 'Genes, as they exist naturally within our bodies, cannot be patented for the simple reason that they have been around for a very long time ... [However], isolated genes that have been removed from the body and copied many, many times constitute something that can potentially be patented if they otherwise meet the criteria for patentability. This is because, in all the eons that have passed since our genes came into existence, they have never come neatly in isolated and purified form. This is one of the hallmarks of an invention: that it would not have existed but for human invention' (Gold 2000, 2–3). Following from the potential of the divisibility of the person/body, isolation is the way in which the divisible parts are rendered amenable to recognition and manipulation. And isolation is a powerful technique. As Patricia Baird notes, with respect to biopatenting law in general, patents are usually allowed for purified and isolated natural substances that exist in nature only in their impure form (1998, 394). Thus, isolation is also a practice of purification, removing the unnecessary, the unwanted, the impurities authored by nature.

From Gold's and Baird's explications of the well-accepted underlying assumptions of biopatenting, it is clear that biopatenting rests upon an equation of genetic isolation with existence – the substitution of visibility for creation. The reason that unisolated genes cannot be patented is that they cannot be seen in their pure form. Jon F. Merz argues, 'Disease patents basically lay claim to an act of observing DNA in the form in which it exists in individuals' cells when such observing is performed for the purpose of diagnosis ... The act of looking itself is not new ... What is new is the empirically discovered association between different isoforms of a gene and some phenotype ... That is, what is new is knowing where to look' (2002, 100). Because the scientist can render the genes visible, the scientist becomes their creator. There is a privileging of the activity of the molecular gaze, within which looking becomes making. An epistemology of visioning (and representation) becomes an ontological activity.

Representation

Related to the isolation, and, hence, visibility of the gene, the third shift necessary for genetic propertization is the representability of genetic substance. This is achieved through DNA's coding as information. Aoki notes the significant role that information plays in the conceptualization of biopatenting: 'DNA is literally the ultimate information replicator and processor, a biological photocopying machine and living embodiment of information's potential infiniteness. Such potentially abundant information would be unavailable to use without the intervention of an inventive human subject' (1993b, 232). The representation of genetic data in the shared language of digital information, which is amenable to computer manipulation (neither a new nor accidental process), enables its human manipulation, and, then, its capacity for invention (e.g., Fox Keller 2000; Kay 2000).

This requirement for rendering in and through language is a long-standing plank of intellectual property's requirements that ideas must be expressed in order to achieve property status. In patent law, the process and the object of the patent must be described fully in the initial application. But even more than this, Rebecca Eisenberg contends that biopatenting seems increasingly to entail information and particular genetic information, stored in a computer-readable medium (2002, 120–1). Representation, given the entangled history of information theory, the modern computer, and genetic science, is thus built into the very conceptualization of the gene.

Authorization

A fourth precondition of the propertization of life is authorization. In this, patents are not unique; all forms of intellectual property require a creator. One of the most enduring myths of intellectual property, and patent law is no exception, is the romantic author. As Aoki (1993b) notes, the author in patent law comprises a combination of scientist/inventor and entrepreneur (see also Etkowitz and Webster 1995). The labour, originality, and entrepreneurialism of an individual, therefore, are imbued into his or her work, with the resulting moral endorsement and economic reward. Following Aoki, we want to explore the implications of the union of the scientist and the entrepreneur – two powerful cultural tropes in the hybrid inventor.

First, the scientist is the agent through whom labour and ingenuity are enacted. Isolated from his or her research team and knowledge community, the individual scientist produces original and creative work that improves society. The applied scientist, in particular, interrupts or patterns the flow of otherwise natural events. 'Themes such as "discovery and invention" turn processual events into localizable facts by fastening them a determinate identity and a stable pattern of causal co-ordinates' (Pottage 1998, 162). At the same time, however, the scientist is also an entrepreneur generating biovalue. The scientist is located within a web of biovalue relations – often through his or her employment by trans-national biotechnology companies, or through his or her research insti-tution's commercialization practices with such corporations. She or he is not expected to share or publish work until its commercial capacities have been explored and legally protected.

These different and, still sometimes at odds, identities – scientist and entrepreneur – unite in the figure of the inventor, the agent of patent law. Patent law rewards this economized scientific practice and yet an-chors it to powerful romantic authorial myths. The inventor, as a figure, authorizes the production of property, imbuing it with a moral element of reward for such socially valuable activity.

Enforcement

Arguably, the forms of genetic material subject to biopatenting do not pre-exist their entry into the relations of biopatenting propertization. Patents are a unique area of property law because they exist only as a re-sult of legislation. Therefore, the requirement of enforceability is the re-quirement for a regime of expertise and authority, which enables the existence, defines the boundaries, and demarcates the effects of patents.

Property has an inextricable relationship with economic value in a capitalist mode of production, and patent law acts to ground and enable that relationship. However, this valuation is not a prerequisite of the biosubject's conceptualization as property. Much biotechnological de-velopment takes place 'on spec,' where a current value has not yet been assigned to the genetic elements. Nonetheless, property rights can and do still accrue. As the Supreme Court minority in the Oncomouse case notes, 'The oncogene is in the genetically modified oncomouse, and it is this important modification that is said to give the oncomouse its com-mercial value, which is what interests the *Patent Act*' (para. 68). Contrary to the Supreme Court justices, Rosemary Coombe argues that property rights do not flow from economic value, but run the other way: 'Market value arises only after property rights have been established and en-forced; the decision to allocate particular property rights is a prior ques-tion of social policy' (1992, 61). Hence, the objects of biopatents be-come things and property first, and only afterward garner a value (possibly). The coupling of property and value is neither determinative nor even necessary.

The activity of patent enforcement is not a neutral adjudication between competing parties in a marketplace. As Hanson notes, 'Patents are not only the gateway to the marketplace, they are the gateway to different ways of understanding and valuing the subject matter of patents' (2002, 164). Patents are, therefore, a codified regime of knowledge with legal authority to enforce existing notions of property, which also legitimate, and are productive of, new ways of thinking about vitality.

With the normalization of these five conceptual shifts in the current historical moment, it is not surprising that the higher life form emerges and is legitimized as property – as biojuridical subject. Biopatents are, therefore, 'technologies of genetic selfhood' (Rabinow and Rose 2003, 492). This emergent biosubject represents a concept of personhood in which the self is divorced from the body as a whole, and does not precede or exceed the elements of the body's constitution. Instead, subjectivity is constituted in the practices of articulating and managing a series of somaticized elements – both material and symbolic. The person emerges, then, through these various labours in the self, in seeking to exploit the bundle of potential rights contained in the body, not only as material substrate, but also as seemingly infinitely reproducible information. It is in the juridification of these elements that the boundaries of the subject emerge. The biojuridical subject, as a bundle of property rights, accrues economic value in some arenas, although not always, and not always for that subject. It requires more than just legislative authorization. It requires active management – legally, scientifically, and financially – to sustain its existence.

Conclusion

While the conceptual shifts producing the subject as property have gone largely unremarked in the Canadian public sphere, it is in popular culture narratives that they are more visible. Two tales of subjects as fungible property in 2005 more effectively contemplate the implications of biojuridical subjectivity than any of the Canadian institutional authorities have done.

Kazuo Ishiguro's powerhouse novel of speculative fiction, *Never Let Me Go*, tells of the boarding school childhood of Ruth, Tommy, and Kathy H. These three young people and their fellow students at Hailsham Estate are clones being raised to serve as multiple organ donors for another class of people, until their ultimate 'completion.' The students of Hailsham are part of a 'socially progressive' experiment where the driven Miss Emily and her colleague Marie-Claude – known to the students as 'Madame' – try to give the children at least a semblance of a normal childhood before they ultimately fulfil their social role as sentient organ banks. In order to secure the support of the broader population, they use the students' artwork to demonstrate that they have souls. By the time that Tommy and Kathy finally learn the truth of Hailsham at the end of the novel, the experiment has failed. Miss Emily explains, 'For a long time, people preferred to believe these organs appeared from nowhere, or at most that they grew in a kind of vacuum. Yet there *were* arguments. But by the time people became concerned about ... about *students*, by the time they came to consider just how you were reared, whether you should have been brought into existence at all, well by then it was too late ... How can you ask a world that has come to regard cancer as curable, how can you ask such a world to put away that cure, to go back to the dark days? There was no going back' (Ishiguro 2005, 240). Miss Emily goes on to describe how her movement battled those perceptions and tried to convince people to see the students as human. But then one scientist attempted to improve the stock, to produce students who were superior.

'After the climate changed, we had no chance. The world didn't want to be reminded how the donation program really worked. They didn't want to think about you students, or about the conditions you were brought up in. In other words, my dears, they wanted you back in the shadows' (Ishiguro, 2005, 242).

Her words echo the less eloquent explanation of human technician McCord (played by Steve Buscemi) in Michael Bay's futuristic thriller, *The Island*, as he explains to Lincoln Six Echo (Ewan McGregor) and Jordan Two Delta (Scarlett Johansson) that they are clones, raised in an idyllic, isolated environment in order to serve as 'insurance policies,' or replacement parts for wealthy people 'out there,' and that their lives are a lie. When he tells a confused Jordan Two Delta that their 'sponsors' actually own her and Lincoln,. she queries why they would not want to know that their clones are alive (rather than kept in the persistent vegetative state that the Institute tells customers they are in). McCord answers, 'Just because people wanna eat the burger, doesn't mean they want to meet the cow.'

Both popular narratives thus address the cultural anxieties of biojuridical subjectivity. The same anxieties are also clearly visible in the Supreme Court of Canada decision in Oncomouse as it clung to an increasingly quaint-sounding humanist notion of the subject. Yet the traditional legal and social categories do not seamlessly fit the animal, human, or plant refigured as a biotechnological hybrid entity. The Oncomouse, like the clones in Ishiguro's and Bay's popular texts, is *already* simultaneously object and subject. By its very existence, it inexorably troubles the fundamental categorization of the higher and lower life form at the heart of the legal and administrative decisions. As Miss Emily says to Tommy and Kathy H., there is no going back.

Life is already an object of biogovernance. In order to bring life itself within the rationality of governmental action, society has rendered it divisible. Contrary to the learned judges, one's sum is often not greater than the total of one's parts. Scientists have already isolated the elements of life; its 'building blocks' are recognizable, and, therefore, it is potentially manipulable. Since Watson and Crick, one can represent that life in code and therefore one can translate, write, and rewrite it. Scientists are recognized as the new, and infinitely more efficient, authors of life, and finally society has adopted a depoliticized legal regime to regulate and enforce all of this activity. This is why, when CBAC writes of 'invented humans,' it is not engaging in speculative fiction. It is speaking the logical and largely normalized outcome of biopatenting as biopolitical technique.

The stakes of biopatenting are high. Who is alive? Who gets to be a higher life form? Who is a who and who is a what? Who gets to own life? The Supreme Court of Canada majority decision in the Oncomouse case, as a last attempt to stem the effects of the five conceptual shifts outlined above, begins as an anachronism. Recently, in another 5–4 decision, which will be discussed in greater detail in the concluding chapter, the Supreme Court of Canada finally ruled on the battle between Saskatchewan farmer Percy Schmeiser and global giant Monsanto, holding that a plant gene, but not the plant itself, was patentable (*Monsanto Canada Inc. v. Schmeiser* 2004). The practical effect of this decision, however, was to allow a patent on a higher life form, albeit one that is non-sentient.

This has thrown the romantic outcome of the Oncomouse case into flux. And, in the absence of parliamentary speech on the biojuridical subject, for better or for worse, these legal decisions are mapping out the governmental terrain for the proper intervention into the conditions of life. They describe the relative powers of the various actors (human and non-human, individual and institutional). They are establishing the truths, in and through which humans, animals, and plants, will and can live. The taxonomies of that problematic category of the natural are rapidly being denaturalized and destabilized in the intersection of law and science. What may seem like legal semantics – what constitutes an invention within the terms of the Canadian *Patent Act* – has significant symbolic and material consequences for mice, canola, and humans as emergent biosubjects.

The social science fictions at the heart of biopatenting emerge not only from cultural texts but, importantly, are mapped out first in legal, governmental, and entrepreneurial terms. In this way, they appear always already circumscribed, controlled by the social actors who seek to benefit the most from the process and who have vested interests in redefining such notions as bodily integrity and sovereignty in ways that allow for the propertization of higher life forms. They have done so with very little public debate through the media or Parliament – the expected arenas for such critical discussions. However, as the brief overview of *Never Let Me Go* and *The Island* shows, popular culture is an important site of rupture into the attempts to seal hermetically public debate along legal and scientific lines. As the previous chapter on reproductive technologies suggests, social science fictions merge the actual and the possible, the pragmatically 'real' with the speculative creations of artists who bring to the forefront the questions that the courts wish to avoid: what are the limits of the body, of humanness altogether, when the self can be broken down into its component parts and granted differing values along a continuum of biotechnological instrumentality?

CHAPTER FIVE

Biosecurity, Bioterrorism, and Epidemics

She scanned every face at the table before answering. 'During the SARS out-break, imagine how easy it would have been to go to Hong Kong, infect yourself, and then intentionally spread it elsewhere.' She paused before turning to the FBI Deputy Director. 'Man-made propagation of a natural epidemic. That, Ms. Roberts, is where I think the terrorists will get the best bang for their buck.'

Kalla (2005, 20)

Prior to 9/11, bioterrorism and epidemic outbreaks were largely fic-tional notions, in the North American context at least. Although the occasional dire warning about the possibility of an intentional release of disease-causing bioagents or the natural eruption of a mutated virus flashed across the media landscape, these were concerns held predomin-antly by fans of thrillers, scientific experts, and political and military au-thorities. One of the earliest widely read fictional representations of an 'outbreak' was Michael Crichton's 1970 bestseller, *The Andromeda Strain*, in which a microbe brought to earth on a space probe begins to infect and kill human beings. The extraterrestrial origin of the germ placed it solidly in the realm of science fiction, reflecting the generalized lack of governmental concern with bioterror and epidemics in a period in which popular and political attention remained largely fixed on the nu-clear threat.

After the end of the Cold War and the lessening of nuclear tension in the 1990s, bioterrorism and epidemics began to inhabit a more central place in the popular imagination. Beginning with Richard Preston's *The Hot Zone* (1994), and followed by other bestselling thrillers such as Preston's follow-up, *The Cobra Event* (1997), Robin Cook's *Vector* (1999),

Douglas Preston and Lincoln Child's *Mount Dragon* (1996), and the film *Outbreak* (1997), representations of deadly disease epidemics began to form a significant proportion of thriller output. With some exceptions, the majority of these narratives focused on the outbreak of naturally occurring epidemics or genetically altered microbes escaping from the research laboratory, rather than their intentional release by bioterrorists. That step in the public imaginary had not yet been made.

With the attack on the World Trade Center on 11 September 2001, and the release of anthrax spores through the United States postal system in the following month, bioterrorism became a central concern around the world. Fiction became social science fiction. Bioterrorism operates as a social science fiction in that it involves its own 'history in advance' (Bogard 1996, 23). It has produced a future that is already a part of the present and is treated as such. Present and future have imploded into an ambiguous present/future that may or may not already be a part of lived experience. In previous chapters we have mapped out social science fictions involving the perfectibility of genetic surveillance, the regulation of non-existent reproductive technologies, and the legal and governmental idea of patenting life. In this chapter we add to the list the social science fictional perception that we increasingly live as though bioterrorism were a normalized condition. Is bioterrorism a real and feasible threat? How is it defined? Has it actually occurred? Are there means of countering it? With the one exception of the small-scale 'Amerithrax' case, there has never really been a significant bioterrorist attack and yet today, around the world, an elaborate biosecurity system is under development in order to counter bioterrorism. The emerging biosecurity system, therefore, is also social science fiction. It is a crude set of interconnections structured around the ideal of becoming a global surveillance network for detecting disease outbreaks and bioterror attacks and countering them before significant loss of life can occur. However, it has never really been put to the test and exists largely in the realm of computer simulations. Neither bioterrorism nor biosecurity has manifested in effective ways, yet both are becoming a part of everyday life, instrumental in shaping the nature of global governance in the twenty-first century.

In this chapter, we examine fiction, public policy statements from the Canadian and United States governments and the United Nations, and press coverage in two widely distributed Canadian newspapers to examine public sphere discourses about bioterrorism and to address a number of questions: How are authorities defining bioterrorism? How are governments seeking to counteract it? What system is emerging to address the potential threats of bioterrorism? What are the broader impacts of this system? How does Canada fit within it? In Canada, the Toronto SARS outbreak in 2003 served as a testing ground for existing response capabilities and also catalyzed Canada's biosecurity establishment. It is an illustrative case study of how several countries around the world are responding to the threat of bioterrorism and epidemics, and what biosecurity means and might mean. We argue that the perceived threat of bioterrorism has initiated a biogovernmental response within Canada and elsewhere that has certain characteristics designed to responsibilize identifiable bioagents, objectify the targets and processes of governance, and normalize a condition of bioterrorism in populations. The resulting biosecurity system, however, has deeper implications than countering the threat of bioterrorism. It is part of an emerging global hierarchy of states, based on the ability of each country to manage the material and symbolic leakiness of its borders.

Defining Bioterrorism and the Condition of Insecurity in the Post-9/11 World

In the pre-9/11 cultural landscape, imagined scenarios of bioterrorism focused on the threat posed by the lone wolf terrorist. For example in Robin Cook's *Vector* (1999), a former Soviet bioweapons expert who has immigrated to the United States is persuaded by a white supremacist group to produce anthrax and botulinum for them. In John Marr and John Baldwin's *The Eleventh Plague* (1998), a rogue scientist decides to use a series of disease agents to infect Christian fundamentalist groups who have worked to cut off his research funding. Mirroring the lone wolf villains are lone wolf investigators, or hastily assembled teams, who must race the clock to piece together the evidence before a major epidemic breaks out. These and other novels and films emphasize the risk of invisible bioagents that are easily and secretly spread. In keeping with the thriller format, the ending is almost always a close call, but is also often indeterminate – the criminal is caught ... but are all of the microbes apprehended? Is it only a matter of time before another outbreak? In this way, pre-9/11 biothrillers portrayed the world as always on the brink of apocalypse, just at a time when global politics were in the process of destabilizing and militarizing. Governments were represented as ineffective, scientists as unstable, and the public as helpless. The fiction gave little comfort; it did not provide an image of biosecurity that could counter the threat of bioterrorism. Often, fictional bioterrorists were caught by luck rather than systematized effort.

Journalistic narratives have contributed further to a sense of insecurity since 9/11. There has been a steady diet of press reports sounding the alarm about epidemics and bioterrorism. For example, prior to the U.S. invasion of Iraq in 2003, the *Globe and Mail* warned, 'When weapons inspectors return to Iraq, their greatest challenge may not be tracking down Saddam Hussein's suspected nuclear and chemical-weapons facilities, but unearthing the country's bioterror labs. Biological weapons facilities are the hardest to detect and the most difficult to shut down' (Sallot 2002). In an article entitled 'Medieval in Manhattan,' the *Toronto Star* reported, 'Thought the bubonic plague was as dead as the millions it took with it in medieval Europe? Surprise. It is not only very much alive and kicking in distant places like India, where the latest scare sent millions running in 1994, but also much closer to home ... It features on the A-list for potential bioterror agents in both the United States and Canada, where it occurs naturally in some of the wildlife population' (Porter 2003).

In addition to the re-emerging threat of old diseases, new ones are apparently appearing. In reference to the avian flu scare one year after the SARS outbreak, the former head of infectious diseases at the World Health Organization (WHO), is quoted as saying, 'We could be in for a very easily transmitted disease and a difficult time ... It would be highly lethal because humans have not been exposed to it before. And the world is not fully prepared for a new and dangerous influenza' (Gerstel 2004). Following the report of a case of mad cow disease in Alberta, academics and union officials were quoted in the *Globe and Mail* insisting that 'Canada needs to enter "a major cycle of reinvestment" in all levels of food safety and inspection to deal with the exploding threat of infectious diseases and bioterrorism' (Leblanc 2003). In another article, the chief of the Department of Microbiology at the Toronto Medical Laboratories and Mount Sinai Hospital sounded the alarm, stating, 'The score of new or re-emerging diseases that have surfaced in epidemics across the globe has shaken the medical community and shattered the commonly held belief that infectious diseases are under control' (Low 2003). Through the use of expert testimony, the threat of epidemics is constituted as credible and imminent.

In recent years, bioterror thrillers and media warnings about bioterrorism and epidemics have continued to proliferate, but a new element of systematization has been added to many of the representations possibly influenced by the CSI effect, discussed in chapter 2. The Canadian television series *Regenesis,* which debuted in 2004, follows a team of scientific experts who work for a joint North American institute called NorBac, which is charged with using hi-tech surveillance systems to track disease outbreaks and bioterror events, investigate them, and determine how to contain them. NorBac is very different from the lone wolf investigators of the past and marks a shift in fictional representation toward a more organized approach to biosecurity. Correspondingly, press accounts have moved to mapping governmental responses rather than representing authorities as ineffective or absent. They are detailing new technological and organizational efforts to prepare the world for the impending era of biological risk. Increasingly, fiction, news media, and government policy produce, and operate within, an emerging discourse of biosecurity.

Before there can be biosecurity, however, there must be an object toward which it can be directed, and this object is bioterrorism. Although natural disease outbreak is far more common than bioterrorism, in the current climate of insecurity and the 'war on terror,' bioterrorism is given a greater level of attention. Discussions of how to manage epidemics are often framed in terms of providing experience for dealing with bioterrorism. However, attempting to define bioterrorism and terrorism more generally has proven to be a complicated task. The High-level Panel on Threats, Challenges and Change, convened by the United Nations, argues for an internationally accepted definition of terrorism as 'any action, in addition to actions already specified by the existing conventions on aspects of terrorism ... that is intended to cause death or serious bodily harm to civilians or non-combatants, when the purpose of such act, by its nature or context, is to intimidate a population, or to compel a Government or an international organization to do or abstain from doing any act' (United Nations 2004, 52).

Legal definitions of terrorism in most nations adopt the spirit of the UN text, but two main objections have been raised that hinder its universal adoption. First, some critics argue that any definition of terrorism should include state use of armed force against civilians. Second, people under foreign occupation must have a recognized right to resistance, which should not be negated by definitions of terrorism (United Nations 2004, 51). Both objections are dismissed by the UN panel, which points out that the UN has already passed normative statements regarding state violence against civilians, and these should be complemented by a normative framework for non-state use of force. In terms of the second objection, the panel argues that the fact of occupation does not justify the targeting and killing of civilians (51–2). Despite the UN's dismissal of these objections, the problem of defining terrorism remains because quite commonly the purpose in describing an act as 'terrorist,' rather than something else, is politically motivated – an attempt to deny it any political legitimacy and to deny any claims that perpetrators are involved in a war of liberation. Therefore, for example, Russian authorities label militant Chechen nationalists as terrorists rather than rebels or guerrillas in an independence movement.

Social science definitions of terrorism have a similar emphasis on civilian targeting. For example, Martin Shaw (2005) in his theorization of new ways of warfare in the twenty-first century defines terrorism as 'a way of using violence designed to cause a political effect by creating terror and fear among a civilian population' (64). He acknowledges that all forms of warfare cause fear and affect civilian populations – this is one of the defining characteristics of late modern warfare – but terrorism involves a principal reliance on generating fear among civilians. This is the primary rationale for terrorist methods; terrorism counts on the spread of knowledge of its violence. Contemporary terrorism is marked by the 'dramatic new lengths' that characterize it and the 'shocking determination' of the terrorists' desire to maximize civilian casualties (65).

Herfried Münkler (2005), in his survey of the 'new wars,' elaborates on Shaw's characterization of contemporary terrorism by distinguishing it from guerrilla warfare. He argues that third world liberation struggles earlier in the twentieth century employed terrorism as a tactical component of guerrilla warfare directed against a selective list of targets that usually did not include anyone belonging to the guerrilla's social, ethnic, or religious group. Further, guerrilla violence generally did not include the use of weapons of mass destruction because it was important to seek legitimacy from third parties. Beginning in the 1970s, religious terrorism became more common and had a different set of motivations, he suggests. It was directed against an absolute evil, had no need for third-party legitimacy, and had less concern for civilian casualties in its struggle against evil. It also internationalized terrorism, taking the attacks to the centres of the powers viewed as oppressive by hijacking airplanes and ships, bombing embassies and tourist sites, and ultimately, destroying the World Trade Center. In other words, terrorist violence shifted from being a *tactic* of guerrilla warfare and became an independent political/ military *strategy*. As a result, limits to the level of violence have expanded and success is now measured by the extent of material damage, the numbers killed and wounded, and, most importantly, the intensity and duration of media coverage (106).

But one can ask of Münkler, against what is this strategy directed? To argue that terrorism is a strategy of political violence is to suggest that there is a terrorist political will that seeks to impose itself and achieve certain ends. Political analysis in the West defines the political will behind much of the terrorism today as based on an Islamist agenda, in keeping with Samuel Huntington's (1997) 'clash of civilizations' thesis, which holds that the dominant fault line in the post–Cold War world is a cultural contestation between Islam and Western liberal capitalism (209). At one level, this appears to be the case, but at another level, many social theorists argue that a broader contestation is occurring – a war against globalization. In this sense, what we are seeing is the chapter that comes after Francis Fukuyama's (1989) bold declaration that, with the end of state socialism in Eastern Europe, we have achieved the global triumph of liberalism and the end of history. According to theorists such as Ulrich Beck (2002), Zygmunt Bauman (2001), and Jean Baudrillard (2002), a new dialectic has emerged and contemporary terrorism, and the corresponding war on terror, is its expression.

Beck (2002) and Bauman (2001) both argue that terrorism is a reaction to the 'neoliberal victory march' and an attempt to undermine the completion of economic globalization. Baudrillard (2002) echoes this view, pointing out that the hegemonically enclosed nature of economic globalization, and its smothering of other forms of social organization, produce an inevitable reaction:

> When global power monopolizes the situation to this extent, when there is such a formidable condensation of all functions in the technocratic machinery, and when no alternative form of thinking is allowed, what other way is there but a *terroristic situational transfer*? It was the system itself which created the objective conditions for this brutal retaliation. By seizing all the cards for itself, it forced the Other to change the rules. And the new rules are fierce ones, because the stakes are fierce. To a system whose very excess of power poses an insoluble challenge, the terrorists respond with a definitive act which is also not susceptible of exchange. Terrorism is the act that restores an irreducible singularity to the heart of a system of generalized exchange. All the singularities (species, individuals, cultures) that have paid with their deaths for the installation of a global circulation governed by a single power are taking their revenge today through this *terroristic situational transfer*. (8–9)

This 'terroristic situational transfer' – changing the rules of the game, identifying specific targets within a generalized system, and inflicting suffering on powerful nations – corresponds to Münkler's description of a shift from tactics to strategies of terrorism. Terrorism is no longer a tactic in a struggle for localized liberation. It is a strategy in a war – the Fourth World War (Baudrillard 2002, 11). All of these theorists are arguing that internationalized terrorism is not primarily about religion or ideology, nor is it an attempt to elicit broad international support. It is a blood sacrifice meant to radicalize the world rather than allow it to achieve the end point of globalization: a unified technocratic system of generalized exchange.

If terrorism is an automatic or inevitable reaction to the powerful forces of globalization, and if terrorism is a now a strategic, internationalized, increasingly devastating weapon intended to draw out media coverage, what are the implications for *bio*terrorism? Arguably, the conditions of possibility for bioterrorism are in place. Drawing upon the United Nations definition of terrorism noted above, we can define bioterrorism as any intentional release of disease-causing bioagents that is intended to cause death or serious bodily harm to civilians or noncombatants, when the purpose of such an act, by its nature or context, is to intimidate a population, or to compel a government or an international organization to do or abstain from doing any act. If it is true that terrorism is oriented toward creating highly dramatic spectacles, large-scale bioterrorism would seem to be a potential source for media images that would haunt generations to come. The vulnerability of target nations would be greatly emphasized, and fear of another bioterrorist act, so easily accomplished, would be extreme. The fear inspired by the spectre of bioterrorism comes from the way it would operate through everyday necessities, spreading through water, food, or air – the three most essential elements of life. No one would be safe and there would be no place to hide. Given the meanings that have been attached to terrorism and the potential disastrous effects of bioterrorism, it becomes imperative to act. Governmental agencies must develop management techniques to contain the risks of bioterrorism. However, bioterrorism is not a self-evident object of governance, and its risks must be defined in such a way that they are identifiable and governable in order to legitimate certain types of governmental responses.

The Problem of Responsibilization

Together, ontological insecurity over disease and bioterrorism, a fear of the consequences of globalization, and an understanding of global conflict as occurring between forces of global economic hegemony and local particularism/parochialism are major components of the conditions of possibility for a new biosecurity system that is emerging around the world. Within this context, two particular actors emerge as significant in the production of biosecurity – scientists and governments. In news media discussions of bioterrorism and in public policy statements from governments and international organizations, scientists and governments are the main targets of responsibilization efforts because they are the agencies involved in the production of biosecurity, but also have a role in the production of bioterrorism. A central question of this period of history is, to what extent can we trust these institutions and these individuals to protect us from the risks of bioterrorism and epidemics and to avoid contributing to the level of risk?

Responsibilizing Scientists

In the 2006 UN report on recommendations for an international counter-terrorism strategy, the secretary-general identifies scientists as important bioagents in bioterrorism, favourably citing the International Centre for Genetic Engineering and Biotechnology's code of conduct for scientists working in the biotechnology field (United Nations General Assembly 2006, 11). The UN report warns, 'Soon, tens of thousands of laboratories worldwide will be operating in a multi-billion dollar industry. Even students working in small laboratories will be able to carry out gene manipulation. The approach to fighting the abuse of biotechnology for terrorist purposes will have more in common with measures against cybercrime than with the work to control nuclear proliferation' (11). The problem is defined as one of individual rogue scientists using biotechnology to produce disease agents in the absence of professional oversight or adequate regulation.

This fear about the responsibility and power of individual scientists is echoed in press coverage of bioterrorism. Illustrative of this issue is the Amerithrax case. In October 2001, shortly after the fall of the twin towers, a bioterrorist sent letters permeated with anthrax through the U.S. mail. The first victim was Robert Stevens, a journalist for the *National Enquirer*. Although Stevens's doctors detected the anthrax, it was too late to save him – anthrax must be treated shortly after exposure. The physicians did alert the Center for Disease Control (CDC), which found anthrax spores in the mail bin at the *Enquirer*. The FBI and the U.S. Army Medical Research Institute of Infectious Diseases (USAMRIID) were notified, and the media warned the public to avoid opening suspicious-looking mail. A secretary in the Hart Senate Office Building then found a letter filled with grey powder and called the police, who again detected anthrax spores. Other contaminated letters directed to NBC, ABC, CBS, and the *New York Post* were intercepted.

The U.S. government went on red alert, and the major investigative agencies met frequently. Unfortunately, communications between USAMRIID and the CDC broke down, with the former framing the incident as a bioweapon attack and the latter framing it as a public health matter (Preston 2002, 172–3). As a result of this misunderstanding, authorities failed to caution the U.S. Post Office, and postal workers began to sicken and die. Antibiotics were then dispersed and the bioterror event was quickly contained, that is, no more letters were sent. Despite the efforts of a large FBI special task force, however, no one has been arrested for the five resulting deaths.

What is interesting from a biogovernmental perspective is the information that came out in the resulting investigation. Evidence pointed to two different sources for the anthrax. First, in February 2002, investigators found that the origin of the anthrax strain used in the attack was a Texas ranch. A cow had died from this particular anthrax virus in 1981. A tissue sample was collected by the local veterinarian and sent to a laboratory at Texas A&M University. From there, a sample was sent to the U.S. Army, which had requested anthrax samples from the lab. Two vials were shipped to Fort Detrick, Maryland – headquarters of the army's biological-warfare research centre. Five years later, two Fort Detrick researchers published a paper reporting that this strain of anthrax was highly lethal. All of the anthrax used in the October 2001 attack was from this particular strain (Recer 2002).

To further confuse the matter, the FBI received an anonymous letter dated 25 September 2001, before the first anthrax case was diagnosed, accusing an Egyptian-born scientist, Dr Ayaad Assaad, of having a vendetta against the U.S. government. Assaad, an American citizen, had worked at Fort Detrick until 1997 and had been involved in 'weaponizing' anthrax. In 2001, he was a senior scientist with the U.S. Environmental Protection Agency. The letter stated, 'Dr Assaad is a potential bioterrorist ... I have worked with Dr Assaad and I heard him say that he has a vendetta against the U.S. government and that if anything happens to him, he told his sons to carry on' (Koring 2002). The FBI cleared Assaad, who stated that he believed that the person who wrote the letter to the FBI and the person who sent the anthrax-laced envelopes, with messages praising Allah and denouncing Americans, were the same person.

Complicating the matter was the fact that many of Assaad's colleagues at Fort Detrick had an 'intense dislike' of him. He was dismissed in 1997 after funding cutbacks, and Assaad sued the government on the basis of age discrimination. In his suit, he detailed an atmosphere in which he and other Arab scientists were subjected to ridicule and racism. Assaad felt that a former colleague had written the letter.

As a consequence of these developments, the press reported a disturbing scenario of home-grown terrorism: 'As the FBI closes in on the anthrax terrorist, now believed to be a scientist at Fort Detrick or one of a handful of civilian contractors who participated in the secret weaponized-anthrax efforts, the circle of suspects numbers only a handful. And the emerging scenario is not that of a botched biological-terrorist attack by al-Qaeda or other terrorists but rather a disgruntled scientist seeking to send a wake-up call to a government that had slashed biological-warfare research' (Koring 2002). Equally disturbing were claims that, although the White House denied finding any suspects, scientists felt that a suspect had been found:

> A leading U.S. expert on biological warfare believes the FBI has identified a prime suspect, and is concerned that no arrest has been made. 'I think I know who it is,' said Barbara Hatch Rosenberg, a microbiologist at the State University of New York who also heads the Federation of American Scientists working group on bioweapons. In an interview, Dr Rosenberg said she believes the FBI identified a prime suspect before she did. Dr Rosenberg and dozens of other scientists were asked by the FBI to assist in narrowing the search by helping identify those who had the expertise, access and possible motive to mail anthrax to the two senators and several media outlets. In an interview, Dr Rosenberg suggested the government might be dragging its feet because of fears that the perpetrator might reveal dark secrets about the extent of U.S. biological-warfare programs. (Koring 2002)

The insinuation here is that the United States, as a signatory to the Biological Weapons Convention that outlaws biological weapons, is likely engaged in covert bioweapons research employing anthrax and is actively involved in covering up a multiple murder of its citizens.

In March 2002, another development in the Amerithrax investigation added more confusion. In June 2001, one of the 9/11 hijackers was treated for a lesion on his leg, claiming that the sore had developed from bumping into something. However, the doctor who treated him agreed with investigators that the sore may have been consistent with anthrax exposure. After reviewing the doctor's records and his treatment of the hijacker, experts at the Johns Hopkins Center for Civilian Biodefense Strategies agreed that anthrax was the most likely diagnosis. Further, the hijacker was rooming with other hijackers in Boca Raton where the first fatal anthrax case occurred. Despite this evidence, the FBI continued to stick to its theory that a lone domestic terrorist was responsible. One thing is clear from the Amerithrax case: bioagents, whether they be microbes, terrorists, scientists, or states, are difficult to contain and difficult to responsibilize. To date no one has been publicly identified as responsible for the Amerithrax attack.

In the aftermath, biologists came under increased media scrutiny: 'A year ago, bioterrorism researchers were seen as dull, or else crazy, plodders on the fringe. Today ... they are the new rocket scientists – rolling in funds, praised as defenders of freedom (when not under suspicion themselves), and likely to offer great medical spinoff benefits to humanity' (Abraham 2002). A common strategy in media accounts at this time seemed to be a public understanding of science exercise reporting on the hard work of responsible scientists and their quest to combat bioterrorism. 'The bioterrorism threat has placed microbiologists on centre stage. In addition to helping to try to track down whoever killed five people with anthrax, microbiologists are being called on to create germ-warfare detectors, rapidly diagnose attacks and create vaccines and drugs' (Keim 2002). Individual scientists and teams of researchers were commonly named and lauded as they searched for vaccines for the Ebola virus (Palmer 2003), discovered something new about how the body fights off infectious diseases (Calamai 2002), or developed quick methods for detecting anthrax spores (McIlroy 2002).

At the same time, any comfort that may have been derived from these accounts of laboratory breakthroughs was simultaneously undermined by reports of scientific irresponsibility. For example, in a series of events that would have been deemed incredible even if penned by Michael Crichton, eleven of the leading microbiologists in the world, many of them involved in developing bioweapons, were found dead between November 2001 and April 2002 (Mitchell, Cooper, and Abraham 2002). One died of a stroke; one disappeared and was later found in the Mississippi River; one was slashed by a sword; one died in a lab accident; two were beaten to death on the street; one shot a colleague and then himself; one was in an airplane crash; one was hit by a car while jogging; and one was found dead of undetermined causes, wedged under a chair, naked from the waist down. No explanations were ever offered other than coincidence.

A few months later, in July 2002, American scientists, funded by the Pentagon, constructed a polio virus from scratch, using inert chemicals that they obtained from a scientific mail-order business (Star Wire Services 2002). The dean of Harvard's School of Public Health was quoted as saying, '"This should really raise some red flags," ... adding that it "means that more complicated viruses can be created – and that it is also possible to create viruses that do not exist in the wild."' The news report goes on to quote J. Craig Venter, the scientist responsible for sequencing the human genome: 'This work should never have been done, funded, or published ... I'd go so far as to say I see it as irresponsible science ... They could have demonstrated their prowess with a (harmless) bacterial virus.'

In the following year, April 2003, the story broke that a South African bioweapons expert offered to sell pathogenic microbes to the CIA. The scientist had participated in a South African bioweapons project in the 1990s that combined genes from an intestinal bacterium with DNA from the pathogen that causes gas gangrene, and had stored some of the microbes in his home. He offered to sell his supply to the American government in return for $5 million and immigration permits for nineteen family members and friends. The Americans declined the offer and reported the incident to South African authorities (Warrick and Mintz 2003).

Stories such as these produce a mixed sense of microbiologists and their motivations. Certainly, scientists are not untouchable and unquestionable – they are a legitimate object of media scrutiny, yet one can see media coverage struggling over how to characterize the place of scientists in the unstable, conflictual world climate of the early twenty-first century. In fact, the Amerithrax story suggests that scientists are shockingly susceptible to political, ethnic, and personal interests, in contrast to the usual representation of the scientist as a dispassionate, depoliticized searcher of objective fact, driven by an inherent ethic of bettering the condition of humanity. Beck (1992) argues that late modernity is characterized by this sense of 'reflexivity' – a growing awareness of, and cynicism towards, the contingency of expert knowledge systems and governmental action. Knowledge is constantly revised and practitioners are often in disagreement, eroding public trust in their expertise. Early modern utopian promises of progress through increasingly certain knowledge and rational control give way to a sense of anxiety as people ponder the scale of potential calamity if the increasingly intertwined globalized, technological, scientific, military, economic, and political systems should fail. Through reflexivity, expert knowledge is exposed to claims for public accountability and loses its extra-political status. Consequently, shortly after the Amerithrax case, news accounts sounded warnings about the scientific community: 'In labs across the United States and Europe, dozens of geneticists are working to create stealthy viruses that can deliver artificially engineered payloads into cells without detection by the immune system ... Scientists joke darkly that it used to take a precocious high school student to make a bioweapon. Today, with the help of pre-packaged kits and automated DNA synthesizers, just about anyone can do it' (Wayt Gibbs 2002).

In this situation, responsibilization begins. In the same month that the anthrax-bearing letters were distributed, the U.S. government passed a law making it a crime to possess biological agents except for research and medical uses. It also required extensive background checks and drug tests on lab workers. However, the law does not go so far as to infringe on privatization. No special licence is required to export DNA synthesizers, sequencers, and other machines employed in microbiological research, prompting the president of the Washington-based Biotechnology Industry Organization to comment, 'Are the safeguards in place appropriate? So far I believe they are ... But are they sufficient? Probably not. I think we need to start thinking now about controlling the availability and export of those types of new instruments that could make it possible for a novice to create a dangerous biological agent' (Wayt Gibbs 2002). Despite fears about scientific irresponsibility, little has been done to limit the potential of scientists to develop bioweapons outside of state regulation. Scientists occupy diverse roles in relation to biosecurity as potential guardians of the public, potential threats to the public, and essential actors in the economic arena. They are double agents – necessary for the production of risks and for their containment. In this situation, the absence of regulation becomes a structuring absence that empowers scientists within the biogovernmental regime.

Responsibilization of Governments

Another important bioagent in the emergent biosecurity system is government, which is intimately linked to scientists researching plague viruses and vaccines as their employer, primary funding agency, or main customer. The inability to trust government accounts of what is happening is another lesson of the Amerithrax case. Is the FBI covering up the fact that it has found the Amerithrax killer? Is the U.S. government involved in developing bioweapons despite treaty obligations not to do so?

How many other governments are involved in bioweapons research and development? Are some governments supporting terrorist attempts to develop and use such weapons? Like scientists, governments are double agents in the management and production of biological risks.

The sense of ambiguity about governments and their role in biosecurity is a product of the history of government involvement in biological warfare. A number of countries engaged in intensive research into biological and chemical warfare after the success of poison gases in the trenches of France in the First World War. The Canadian government established its own 'germ' warfare project in 1940 under the direction of Frederick Banting, headquartered at the University of Toronto with testing sites at Grosse Île, Quebec, and near Medicine Hat, Alberta (Regis 1999, 69–70). After the Second World War, Canada ceased its involvement in bioweapons research, but in the United States, major bioweapon research began in earnest in the early Cold War era, centred at Camp Detrick, which had been established for biological warfare research during the Second World War. Employing scientific research taken from the Japanese after the war, scientists at Camp Detrick developed seven types of weaponized microbes that can cause illnesses including anthrax, botulinum, tularemia, brucellosis, equine encephalitis, staphylococcus, and riketts (Drexler 2002, 248).

In 1972, President Nixon signed the United Nations Convention on the Prohibition of the Development, Production and Stockpiling of Bacteriological (Biological) and Toxin Weapons, and on Their Destruction (the Biological Weapons Convention), presumably ending the American biological weapons program. Apparently, the government of the Soviet Union did not believe the Americans had ceased research and development in this field and, despite also being a signatory to the convention, continued its bioweapons program, named *Biopreparat*. According to defecting Soviet scientists, *Biopreparat* had developed vaccine-resistant strains of anthrax, tularemia, smallpox, and black plague and had experimented with Ebola and Marburg viruses as well (Alibek 1999). Defectors also reported that there had been a laboratory accident in 1979 resulting in an anthrax epidemic in the city of Sverdlovsk. Western observers had been told at the time that tainted meat had been the cause of the illnesses (75). With the collapse of the Soviet Union in the early 1990s, the Russian government claimed to have closed *Biopreparat*, although, to date, this assertion is unconfirmed. A number of Soviet scientists are also believed to have moved to 'rogue' states such as Iran, Iraq, and North Korea to participate in biological weapons programs (Drexler 2002, 253). The extent to which this is true is un-known, at least outside of intelligence circles, but its circulation as al-legation in the public domain continues to produce the 'culture of ru-mour' that supports calls for a biosecurity regime.

Enabled by the fearful climate after 9/11 and the ensuing anthrax at-tacks, we seem to be entering a new era of bioweapons research. News reports of military experiments with genetically altered forms of anthrax in the United States and the construction of a synthetic polio virus from inert matter, funded by the Pentagon, has 'reinforced suspicions that the United States may be involved with biological weapons to counter the threat from "rogue" states such as Iraq, Iran and North Korea, all of which have proven bioweapons capability' (Avery 2002). The United States has also resisted attempts, begun in the early 1990s, to develop an international protocol for declaring biological weapons and inspecting them, arguing that the proposal threatens U.S. commercial proprietary information and the Pentagon's biodefence planning. It also argues that the protocol is not intrusive enough to detect the biological weapons programs of 'rogue' states (Avery 2002). Given the history of bioweapons research, the supposed shutdown of these research programs by major states, the ongoing hints that the programs are still secretly operating, and accusations that enemy states are vigorously pursuing bioweapons research, state involvement in bioweapons research becomes very much a social science fiction in which the actual state of affairs is largely un-known, perhaps even to nation states themselves. There is a great desire to know the extent of the danger the world faces and we must live, the logic dictates, as though bioweapons are a part of our concrete reality even as we wonder whether they really are as widespread as we fear. As in bioterror thrillers, any conclusions we can reach are indeterminate.

At the international level, the United Nations and the World Health Organization have taken the lead in defining the problems of bioterror-ism and epidemics and initiating global strategies for responsibilizing state actors. In his April 2006 report to the General Assembly of the United Nations, Secretary-General Kofi Anan summarized much of the accumulated wisdom about the nature of terrorism and the resources upon which it relies: 'Terrorists require means to carry out their attacks. The ability to generate and move finances, to acquire weapons, to re-cruit and train cadres, and to communicate, particularly through use of the Internet, are all essential to terrorists. They seek easy access to their intended targets and increasingly look for greater impact – both in numbers killed and in media exposure. Denying them access to these means and targets can help to prevent future attacks' (United Nations 2006, 8). Based upon counterterrorist work that has been done, especially since 9/11, the report outlines the target areas in which governments can reduce the effectiveness of terrorists, placing special emphasis on biological weapons because of the difficulties in detecting them: 'The most important under-addressed threat relating to terrorism, and one which acutely requires new thinking on the part of the international community, is that of terrorists using a biological weapon. Biotechnology, like computer technology, has developed exponentially. Such advances herald promising breakthroughs and are one of the key battlefronts in our attempts to eliminate the infectious diseases that kill upwards of 14 million people every year. They can, however, also bring incalculable harm if put to destructive use by those who seek to develop designer diseases and pathogens' (11).[1] There is a Biological and Toxin Weapons Convention (BWC) in place, first drafted in 1972, with 171 signatories, as of June 2005. Of the twenty-three nations who have not signed the agreement, noteworthy examples include Israel, Kazakhstan, Kyrgyzstan, Tajikistan, and several African countries. Egypt, Somalia, Syria, United Arab Emirates, and several other countries have signed the treaty but not yet ratified it. All of these countries are defined by Western governments as located in actual or potential zones of conflict.

Article I of the BWC identifies biological agents that each signatory 'undertakes never in any circumstances to develop, produce, stockpile or otherwise acquire or retain.' The nature of these items is broadly defined, not as biological weapons, but as '(1) Microbial or other biological agents, or toxins whatever their origin or method of production, of types and in quantities that have no justification for prophylactic, protective or other peaceful purposes; (2) Weapons, equipment or means of delivery designed to use such agents or toxins for hostile purposes or in armed conflict.' The wording of these provisions is specifically designed to avoid obstructing biomedical research and development. Article II of the convention requires states to destroy or divert to peaceful purposes any existing stockpiles of microbial agents, toxins, weapons, equipment, and means of delivery, within nine months of ratification.

Every five years, a review conference is held at which signatories to the BWC meet to examine the operation of the treaty and discuss ways of strengthening it. It is admittedly a weak instrument; it is only four pages long and contains no formal procedures for monitoring compliance or investigating countries suspected of violating the treaty. This can be compared to the 1993 Chemical Weapons Convention, which is over two hundred pages long, has elaborate verification procedures, and is overseen by the Organisation for the Prohibition of Chemical Weapons. The sixth BWC conference was held in December 2006 and, despite tensions between the United States and Iran, was considered a modest success. Although no major developments came out of the conference in terms of strengthening the verification procedures, there was consensus to have annual meetings of member states and experts to exchange scientific, technical, and political information in order to keep dialogue open in the lead up to the seventh conference in 2011 (Sixth Review Conference 2006).

Because there is no strong international regulation of bioweapon development, the World Health Organization (WHO) has stepped in, taking the initiative to mobilize its member states to coordinate their preparation for bioweapons attacks and epidemics. They argue that the developing international regime of bioweapons response has a strong public health orientation. In its 2002 report, the WHO set out its strategy for managing bioterrorism, which includes 'collaboration of the intelligence community, law enforcement agencies, public health professionals, and the biomedical sciences' (WHO 2002, 2). The document begins by clearly stating WHO's role in the case of a bioterrorist attack. It emphasizes WHO's political neutrality and its sole concern with public health. It also states that, given the difficulties in pre-detecting and pre-empting bioterrorist attacks, the key strategy must be preparedness through strong public health infrastructures.

In order to determine the occurrence of an outbreak, natural or deliberate, the local public health system must be able to chart the geographical, demographic, and clinical features of the disease. Such surveillance must occur globally to ensure that epidemics can be quickly detected and contained. Every outbreak is to be treated as natural until proved otherwise. The first priority of the health system is to reduce mortality and prevent spread. If the outbreak is a covert release of a bioagent, it will likely take days, if not weeks, to verify it, since suspicion of an attack will emerge only when patients begin appearing in health care facilities with unusual symptoms or inexplicable diseases. Therefore, the first line of defence consists of health care workers, who should be given priority for protective equipment and vaccines, and who should be trained to recognize and manage the illnesses that are likely to be used in a bioterrorist attack. They must also be trained in techniques of 'barrier nursing' – safe handling of samples, and decontamination procedures. A major concern for public health systems is their surge capacity to handle a sudden wave of epidemic victims.

WHO provides advice to governments on how their public health infrastructures should prepare for the risk of bioterrorism. It suggests that governments should not establish specialized bioterrorism response units because the infrequency of call-out could lead to a deterioration of skills, the expense of maintaining these teams would be difficult to justify in many countries, and excessive centralization could increase response time. Emphasis should be placed on supporting existing emergency response and public health services at the local level. Specialists are needed, however, for sampling and analysis. WHO oversees a global network of 270 collaborating institutes and laboratories. There is no need to build more of these facilities in more countries. WHO also does not recommend stockpiling drugs and vaccines unless a threat seems 'possible, probable, and specific' (2002, 8). The reasons are the expiry of vaccines, limited global vaccine production, and, most importantly, the diversion of resources from responding to existing infectious diseases, many of which are preventable but continue to spread for lack of access to drugs and vaccines. Despite these recommendations, governments in developed nations have done largely the opposite by establishing emergency response teams, stockpiling vaccines, and reinforcing their own bioresearch facilities.

In its report, WHO recognizes that 'no country can ever guarantee the total security of its population against a biological attack, especially when a contagious agent is used. As with naturally caused outbreaks, the harm is delivered by invisible, highly mobile, microscopic agents that easily cross borders, placing all countries at risk. The consequences – whether in the form of cases of disease or waves of panic – can quickly spread in a highly mobile, interconnected, and electronically linked world' (10). In response to the global risks of bioterrorism and epidemic outbreak, WHO has developed the Global Outbreak Alert and Response Network (GOARN), which formed in 1997 and was formalized in 2000. It is an interlink among over one hundred existing data-gathering networks, including governments, universities, ministries of health, academic institutions, other UN agencies, networks of military laboratories, and non-governmental organizations, which keeps track of reports of disease outbreaks around the world. It employs a computer tool, the Global Public Health Intelligence Network (GPHIN), which constantly scans websites, news wires, local online newspapers, public health email services, and electronic discussion groups for rumours of outbreaks. In this way, GOARN hopes to keep track of informal information that may warn of an unusual disease event. GPHIN was developed by Health

Canada and is managed by the Canadian Public Health Agency on behalf of WHO.

Despite an international and domestic rhetoric of the need to responsibilize scientists and governments, little has actually been accomplished since 2001. The strongest response has been the beginnings of an international surveillance system to detect disease outbreaks and to mobilize international laboratories. UN attempts to strengthen the Biological and Toxin Weapons Convention to reduce the threat of state and non-state development and use of bioweapons have been ineffective. Regulation of scientific activity founders on the need to ensure that bioindustry is not unduly hindered. There is no international effort to build up public health systems in developing nations. Fear of terrorism and the general permeability of borders to the flow of scientific expertise, technologies, and commercial products undermine attempts to responsibilize science. While the press reports on scientific projects oriented toward the public good, it concurrently unravels reassuring perceptions with stories of leaky borders that threaten developments in scientific knowledge, and hints that governments are involved in bioweapons research and development. With the limited success of responsibilization strategies, the resulting sense of insecurity contributes to the formation of a biosecurity system that promises to counter the threat of bioterrorism and epidemics through a systematized linking of national and international organizations and resources. It is just such a system that is under development in Canada and around the post-industrial world.

National Biosecurity Strategies

Given the poor track record of governments in refraining from bioweapons research, in managing scientific responsibility, and in controlling the proliferation of relevant technologies, governments must act in some way to maintain legitimacy for the management of the risks of bioterrorism and epidemics. As a result, since 9/11, governments and international organizations have engaged in a flurry of biosecurity activities that fall into the following three proposed categories: vaccine research and stockpiling, research into microbe detection and defence technologies, and bureaucratic centralization of bioterrorism preparedness.

The primary technical response by governments has been to stockpile vaccines and invest in research into new vaccines. Accompanying doom-laden press reports about the advent of new epidemics and our lack of preparedness are numerous stories about advances in vaccine research.

Among them are stories about breakthroughs in vaccines for the Marburg virus, successfully tested on monkeys; West Nile virus, with research at the earliest stage; Ebola virus, a vaccine for which has also been successfully tested on monkeys; and anthrax, a vaccine for which is under development by pharmaceutical company VaxGen Inc. (Carey 2005; Frketich 2004; Talaga 2005; Tobin 2006).

Most attention, however, has been focused on the smallpox vaccine. In November 2002, Health Canada contracted with Aventis Pasteur Ltd of Toronto to supply ten million doses of smallpox vaccine as protection against a bioterror attack (Harper 2002). Smallpox is a major concern largely due to Western intelligence information about a Soviet and ongoing post-Soviet program in Russia to develop smallpox as a biological weapon. Consequently, U.S. troops in risky areas are routinely vaccinated for smallpox and the United States has a stockpile of 300 million doses of smallpox vaccine for a nationwide inoculation program, if it is deemed necessary (Shephard 2004).[2]

One problem that must be faced by governments in their pursuit of defences against bioterrorism and epidemics is a generalized public distrust of government and science and the fact that vaccines have become a form of contested knowledge in recent years. Ontological insecurity is directed not only against bioterrorism, but also, to some extent, the potential solutions to bioterrorism and epidemics. A survey of 1,000 Canadians taken in January 2002, conducted by Cancer Care Ontario, found that only 57 per cent of respondents would take a vaccine recommended by the government. Fifty-two per cent felt that vaccine safeguards are 'slack' and 40 per cent disagreed that vaccines are medically effective. Forty-six per cent stated that the idea of taking a newly developed vaccine, even if carefully tested, made them anxious. Perhaps most surprising was the fact that 33 per cent did not know what vaccines were and why they received them as children (Yelaja 2002). Canadian scientists expressed extreme puzzlement over these findings and argued that 'the public needs some serious education on the benefits of vaccination' (Yelaja 2002). One scientist pointed out, 'Vaccines are a highly effective treatment. But there's a big knowledge gap for a large proportion of the population, which could hinder our ability to intervene quickly in the event of a bioterrorism event ... This is all unprecedented for Canada. We've never had situations of immediate danger like that, so we need to prepare' (Yelaja 2002). Scientists interpret this public response, predictably, not as a problem of their own lack of responsibilization to the public but as a matter of a poorly educated and misinformed public.

They seem to have little sense of the ontological insecurity produced by modern science and the generalized reflexivity toward scientific expertise that is characteristic of late modern society. Vaccines have been a subject of public debate among parents for a number of years, but scientists do not seem to recognize the contested nature of their knowledge.

In addition to vaccine research and stockpiling, other technological measures have been taken in Canada and elsewhere. On 11 July 2002, the Ontario minister of public safety announced the purchase of 3,800 bioterror suits to be distributed to Ontario police officers. The gear consists of a full-length spacesuit-like body suit, butyl rubber gloves, Tyvek boots, a black respirator mask, and a biohazard disposal bag, all to be carried in the trunks of police cruisers (Canadian Press 2002). Research into sensors to detect microbes is also ongoing, primarily in the United States. Although Iraqi forces did not use biological or chemical weapons, the 2003 invasion of Iraq was a testing ground for this type of technology, managed by the U.S. Army Chemical School and funded by the U.S. Defence Advanced Research Projects Agency (DARPA). The commandant of the Chemical School stated in 2002, 'We were the science fiction of the military, because people thought that something would never happen in this area [bioterrorism] ... But like most science fiction, it's now become science fact' (Ross 2002). Although existing sensors are slow, bulky, expensive, and have a tendency to give many false alarms, they have been established in subway stations in Washington, D.C., and New York, as well as the San Francisco International Airport. Their value in warning of an impending attack is currently limited by the fact that air samples collected in the sensors must be sent to laboratories for analysis, delaying discovery of an attack by as much as twenty-four hours. The value of this type of sensor is the identification of a disease microbe before it has time to fully incubate in a human host, allowing for treatment of affected victims (Chang 2003).

In addition to technological solutions, governments have engaged in bureaucratic centralization as a means of combating terrorism. The most famous example is the U.S. Department of Homeland Security, which was legislated into existence on 19 November 2002. Homeland Security ambitiously combines twenty-two government agencies – each with its own organizational history, mission, and authority – into one organizational framework, including the Coast Guard, Immigration, Border Patrol, Customs, the Secret Service, the Federal Emergency Management Agency, and other emergency response agencies. It has approximately 180,000 employees and a budget of $38 billion (Kenney 2007, 176). Its mission is very broad: to protect the United States from terrorist attacks. Under pressure from the urgency of its mandate, the complexity of restructuring a large number of government organizations into one organizational culture, and the difficulties in training personnel to deal with catastrophic terrorist attacks despite the problem of a small sample from which to learn, Homeland Security is struggling to develop counter-terrorist security strategies, as well as new means for monitoring population flow and population surveillance, in the face of its complex bureaucratic processes (179).

In terms of legislation, the United States has a number of statutes that regulate responses to bioterrorism, including the 1989 *Biological Weapons Anti-Terrorism Act* and the 1991 *Chemical and Biological Weapons Control and Warfare Elimination Act*. After 9/11, further legislation was passed. As a part of the *Uniting and Strengthening America by Providing Appropriate Tools Required to Intercept and Obstruct Terrorism* Act (USA *Patriot Act*) of 2001, the biological weapons legal arsenal was expanded to include additional offences related to possessing and transporting biological agents and delivery systems, as well as expanding the amount of funding invested into countering bioterrorism. In June 2002, the *Public Health Security and Bioterrorism Preparedness and Response Act* was passed, detailing the distribution of responsibilities for bioterrorism preparedness and response and setting out an agenda for strengthening biodefence standards and protocols, as well as police powers in the apprehension of bioterrorists.

In Canada, similar centralization is underway. Key to Canada's preparation for a bioterrorist attack is the Department of Public Safety, created in 2003 as a response to security concerns after 9/11. Like Homeland Security in the United States, Public Safety is an attempt to ensure that there are clear lines of authority and communication in times of emergency. The *Public Safety Act* (2002) links crime management with national security under one portfolio by bringing together the Canada Border Services Agency, Canadian Security Intelligence Service (CSIS), Correctional Service Canada, the Royal Canadian Mounted Police, and the National Parole Board under the direction of one minister. In a national emergency, this minister will be the authority in charge of coordinating a response. The minister of public safety is also the chief minister responsible for enforcing the *Anti-terrorism Act* (2001), Canada's primary legislative tool for apprehending suspected terrorists and disrupting their operations in Canada. Although the *Anti-terrorism Act* does not specifically address bioterrorism, its definition of terrorism clearly applies to bioterrorist acts (see section 83.01(b)).

Key to responding to the effects of a bioterrorist attack, in particular, is the Public Health Agency of Canada (PHAC), which was created in 2004 in the wake of the Toronto SARS crisis. In Canada, municipal and provincial governments have their own emergency response systems, but their resources are limited. PHAC is available if assistance is required locally or if an emergency crosses provincial borders. In terms of bioterrorism and epidemic outbreak, PHAC has a number of responsibilities, including developing and maintaining national emergency response plans; managing the quarantine service at border crossings; developing laboratory protocols for testing potential biologic terrorism agents; training lab workers in their use; maintaining and deploying mobile laboratories and emergency response teams anywhere in Canada or abroad; transporting high-risk microbes; monitoring disease outbreaks through the Global Public Health Intelligence Network (which it also manages on behalf of WHO); managing the government stockpile of vaccines and emergency medical supplies; providing equipment and training to provincial hospital workers; and providing for emergency medical response surge capacity by having emergency response teams ready to deploy to hospitals in the case of medical disasters. The Public Health Agency is headquartered in Winnipeg, with offices in Ottawa. Winnipeg was chosen because of its National Microbiology Laboratory, one of fourteen Level IV testing laboratories in the world (these laboratories handle the most dangerous, incurable disease agents).

In addition to the services it would provide in the case of an outbreak, PHAC participates with the Department of National Defence and the Department of Public Safety to provide Emergency Operation Centres to crisis points within Canada. These are mobile communications centres employing satellite and cellular technology, smart boards and writable walls, media monitoring, telephones with visual display, emergency management software, and other up-to-date communications technologies to help manage any type of emergency. In the event of a bioterrorist attack, the minister of public safety would coordinate emergency management under the authority of the *Emergency Management Act* (2007), by mobilizing emergency operation centres that would include PHAC medical teams, military personnel, and any other forms of expertise required for addressing the emergency.

With the formation of the Department of Public Safety and the Public Health Agency of Canada, we can see a centralizing and normalizing of biosecurity. Prior to 2003, Canada had no government agencies dedicated to detecting and reacting to bioterrorism and epidemics in the same way. The formation of these bureaucracies announces that a new era of bioterrorism and biosecurity is here to stay and has become a normal element of twenty-first-century life for which we must always be prepared. It institutionalizes the governance of bioterrorism. As well, the system represents the spread of securitization into the realm of health, transforming the planning for epidemics into an issue of health security rather than public health. By framing the issue as one of security, health can be linked to the broader convergence of police and military security that is represented by the combination of policing, corrections, border patrol, and intelligence services under one government department.

Biosecurity and the Normalization of Bioterror

What is interesting about development of the Department of Public Safety and the Public Health Agency in Canada, as well as intensive vaccine research, vaccine stockpiling, police bioterror suits, and disease microbe detectors is the fact that none of these pre-existed 9/11 or, at least, not nearly to the same magnitude they have already achieved in a relatively short time. The same can be said for global developments; global outbreak surveillance has accelerated very rapidly since 2001. At the same time, the economic powers of the world are involved in open warfare and occupation of other countries in a way that they have not been since the Second World War. These are the organizational, political, military, scientific, and technological elements, we suggest, of an emerging global biosecurity system. However, it is important to step back and analyse the question of what sort of security is promised by this system and how is it defined.

Mark Neocleous (2000) argues that we have entered an age of insecurity where increasingly the concept of security and its opposite, insecurity, have come to be defining terms for the hazards and problems of late modern society (7). The turn toward 'securitizing' social issues and expanding the scope of the concept beyond its traditional use in state and military terms is a product of expert analysts and influential political figures of the past two decades. In the 1990s, the Clinton administration, the Yeltsin administration, and UN Secretary-General Boutros Boutros-Ghali all called for replacing narrower notions of national security with a notion of 'human security,' which incorporates the categories of health, food, economic, environmental, personal, community, and political security (Neocleous 2000).

An expansion of the language and ideology of security into these categories reflects the general level of insecurity of the late twentieth and early twenty-first centuries. In some ways, such a shift is comforting but is also problematic, because it essentially depoliticizes social issues. Security is such a powerful and colonizing discourse that it is virtually impossible to argue against, with the result that alternative discourses such as the public good, bodily integrity, privacy, due process, and public health are seen as soft and become expendable. At the same time, in the contemporary state of insecurity, it seems very easy to employ a discourse of security to legitimate governance through emergency powers, executive directives, secrecy, and public relations campaigns (Agamben 2000, 113).

Traditional usage of the concept of security derives from international relations studies, which frames it in terms of political realism – the idea that in the state-based international system, states are motivated by the need to preserve and expand self-interest. Having its origins in ancient Greek thought, this dominant security narrative carries on today through the work of contemporary theorists who argue that security has as its final goal the measurement and control of all objective threats (Walzer 1977). In this way security, as an analytical category, is defined as an object that can stand as a pure, idealized signifier in its own right. It is ahistorical and has no constitutive relations or authorized subject position – it simply exists as an object, free of historical social relations. Consequently, the Canadian *Public Safety Act* and the American *Patriot Act* can claim to override classical legal norms in favour of apparently normless and non-politicized, security-based, executive decrees.

If security is an object, it can be known with ever-greater certainty through positivistic exploration of its limits and boundaries, which leads to the emergence of authorized security experts who are identifying gaps in security knowledge. These gaps are the differences between security measures and the targets they seek to control. Security, from this point of view, is expanded as targets are objectified, quantified, commodified, and, ultimately brought under control. With the recent and ongoing growth in security expertise, the number of targets of security can be expected to grow as experts seek out and process a steady stream of threats and insecurities that must be brought into the realm of security (de Lint and Virta 2004, 472).

After 9/11, the pre-eminent security target in the world has been terrorism, prompting a number of states, led by the United States, to embark on a 'war on terror.' The Canadian government has been an active participant in the war on terror, sending troops to Afghanistan as part of the NATO force that invaded that country in 2002 to oust the Taliban regime and destroy al-Qaeda bases for terrorist training. However, Canada declined to participate in the 2003 invasion of Iraq for lack of evidence of terrorist bases or weapons of mass destruction. Canadian forces maintain a high presence in Afghanistan, sustaining relatively high casualty rates on behalf of the allied mission. Employing military force in other countries in order to disrupt the build-up of terrorist groups is an important part of Canada's post-9/11 national security policy and that of many other Western nations as well (Privy Council Office 2004, 37).

From a critical perspective, the linking of security discourse to the war on terror produces what Michael Hardt and Antonio Negri (2000) refer to as a process of 'empire' – a global system in which the sovereign authority of nation states is declining in favour of a globalized system of political and economic organization. This empire is not centred on any one nation-state but is a network of nation-states, international agencies, non-governmental organizations, and transnational corporations that together form a multi-level, non-territorial decision-making network. It is strategic in the sense that it has an agenda of social transformation, or 'liberal peace,' to produce global security, or at least security within the borders of the empire (Duffield 2001, 11).

The problem with producing an empire based on the concept of security is that it requires the constant exercise of force and, ultimately, war. Hardt and Negri (2005) argue that war and security have become the new bases for biopower after 9/11 in the sense that the security needs of the war on terror are beginning to produce the organizing principle for a growing number of aspects of social life. A war against a concept or a set of practices such as terrorism is theoretically endless in that it has no definite spatial or temporal boundaries. It can extend anywhere for any period of time. It is a war to create global social order and to do so must involve the continuous, uninterrupted exercise of power and violence. However, one cannot win such a war in a final sense; it must be won all over again every day. As a result, from a governmental perspective, war becomes more and more like domestic crime control, which must also be fought day by day. International relations and domestic politics begin to converge around the notion of security. Techniques and technologies of military action and police action begin to converge into a global security system, which, in turn, increasingly relies upon a depoliticized state of exception to ensure that there are no procedural hindrances to state violence if it is deemed necessary (14). Contrary to Baudrillard's argument that contemporary terrorism disrupts the emerging global order, the war on terror, arguably, reinforces the global order and the authority of political, military, and economic hierarchies, through a discourse of security in the face of war.

SARS and Canada's Place in the Global Biosecurity System

Production of a security empire, the convergence of military and police security at a global level, and an ongoing and perpetual war on terror, combined with a generalized ontological insecurity over terrorism, bioterrorism, epidemics, vaccines, and the failure of scientific and government responsibilization strategies form the conditions of possibility for the emerging biosecurity regime forming in Canada and around the world. It was in this context that the first major post-9/11 epidemic event occurred in 2003 – the SARS outbreak in Canada and Southeast Asia. SARS forms an interesting case study of the interaction between post-9/11 apparatuses of security and insecurity, the operation of the global biosecurity regime, and its implications for the positioning of nations in the new security empire.

Reports of SARS in Southeast Asia began to filter into the Western media in early 2003. SARS is the acronym for severe acute respiratory syndrome and its symptoms include breathing problems, dry cough, and flu-like symptoms such as muscle aches, headache, sore throat, and high fever. It has an incubation period of two to seven days and is more severe than the average flu, with a greater chance of killing its victims. By early March, WHO reported over 150 cases of the illness in China, Indonesia, the Philippines, Singapore, Thailand, and Vietnam. On 12 March WHO released its first ever 'emergency travel advisory' to slow the spread of the disease. For Canada, however, it was too late. An elderly woman from Toronto, who had recently visited Hong Kong, died of the disease on 5 March. A few days later, her son went to the Scarborough Hospital complaining of severe flu-like symptoms. He died on 13 March after exposing nearby patients and health workers to the disease.

From there, the disease spread and Scarborough hospital was forced to shut its doors on 23 March. Patients were transferred to other hospitals in the region, which were soon overwhelmed by the number of cases. On 25 March, the Ontario Government designated SARS a 'reportable, communicable, and virulent disease,' under the *Health Protection and Promotion Act*, allowing public health authorities to track infected people and quarantine them in order to stop the spread of the infection. The next day, Premier Ernie Eves declared a provincial emergency, which included the activation of 'code orange' in Toronto area hospitals, meaning that all non-essential services ceased, visitors were limited, isolation units were established, and protective clothing was required by staff. Code orange was later extended to all Ontario hospitals. By this point, nineteen people had died (eighteen in Ontario and one unrelated case in British Columbia), and before the containment of the outbreak in May, 44 people died and 375 others were infected (SARS Commission 2006, 28).

Eventually, the primary news story about SARS was not the deaths it was causing or its impact on the health system, but rather the fact that on 23 April WHO issued a travel advisory for Toronto, warning travellers to stay away unless their travel was essential. The advisory was lifted on 14 May, but the economic effects on Toronto were devastating. A study by the Canadian Tourism Commission estimated that SARS cost the Canadian economy $519 million in 2003 alone and estimated an additional $200 million loss over the following three years until Toronto was deemed a safe travel destination again (CBC News Online 2003). Lifting the travel advisory did not immediately end the symbolic linkage of Toronto and death. In order to lure people back to the city, hotels, airlines, and restaurants drastically lowered their prices, yet even cut-rate prices did not revitalize the tourism industries. In a desperate measure, Toronto officials decided to ask the Rolling Stones to headline a benefit concert, SARSstock, to demonstrate the city's safety. On 30 July close to 450,000 people attended the concert, marking a sense that SARS was finally under control and Toronto was beginning to recover economically. The symbolic and political effects of SARS, however, continued to be felt. By early April, three weeks after the Toronto outbreak was first identified, the press had already begun to examine its wider social meanings: 'The fact is that the utter havoc it's wreaking on the health system, on our lives and the life of our city, on international travel and commerce, may be far more serious than the disease itself. SARS itself isn't even the crisis any more. The crisis is the crisis' (Gerstel 2003a).

It is instructive to analyse press accounts and the policy document produced by the federal SARS commission during and after the crisis to examine the meanings that came to be attached to the disease in Canada. These texts indicate an already present set of public fears and values that support the emerging biosecurity regime with its goal of ensuring that Canada keeps pace with international biosecurity developments.

Early on in the SARS outbreak, media reports were using the SARS case to ask whether or not Canada was adequately prepared for a bio-terror attack (Shephard 2003). Finding that it was not prepared, reporters then began to question the leadership of municipal, provincial, and federal levels of government, asking 'why federal, provincial and civic leaders had little or nothing to say about the SARS crisis for five weeks while this city was becoming the pariah of the continent – and now the world; while we in Toronto have been suffering, if not from SARS, from the fallout of fear and confusion' (Gerstel 2003b). The report goes on to quote a number of readers who sent in comments stating their lack of confidence in political leadership and, while starting as a critique of the handling of the SARS crisis, the article becomes a criticism of the state of public health care in Canada, an often-repeated theme in press coverage of the period. '"SARS was an accident waiting to happen – because of the priorities of the government, the cost-cutting measures, the conditions were great for SARS to take hold," said William Bowie, an infectious disease specialist ... People on the front lines fighting SARS say it is nothing short of a miracle that a "bare-bones" public-health system managed to control the crisis ... "It's been very clear to us that we were going to pay for the public-health dismantling that has happened under the provincial and municipal governments," said Allison McGreer, head of infection control at Toronto's Mount Sinai Hospital' (Abraham and Priest 2003).

These and many other media commentaries reflected a generalized sense of frustration in Canada over developments in public health care over the previous two decades and a readiness to blame a crisis such as SARS on neo-liberal cost-cutting rhetoric and policies that have never been popular. The federal SARS commission convened to evaluate Canada's ability to respond to public health crises. The commission's report emphasized that 'while the public health and health care workers involved did an admirable job of containing SARS and keeping it from spreading to the larger community, the SARS experience highlighted weaknesses in Canada's public health system. Many issues to do with the clinical system and clinical/public health interface were also thrown into high relief. Aside from the lack of surge capacity to deal with this crisis situation, problems emerged with respect to timely access to laboratory results, information sharing, data ownership, and epidemiologic investigation of the outbreak. Communication to the public was sometimes inconsistent, and it was not always clear who was in charge of the outbreak response' (National Advisory Committee on SARS 2003, 20). The SARS experience illustrated that Canada is not adequately prepared to deal with a true pandemic and had not bothered to invest in a health crisis management program in the post-9/11 era. In the view of the press, and the committee, this was another indication of governmental irresponsibility.

Lack of government planning seemed even more irresponsible, given the increasing level of global population movement and population growth. A second prominent media and policy theme that emerged from the SARS crisis expressed anxiety about invasion from outside within a globalizing world: 'Certainly, our fears about SARS speak to a fear of interpenetrability – the speed with which an extremely infectious virus can travel around the world ... SARS attacks quickly. We see the SARS danger as a nasty side effect of contemporary globalization. The speed of its journey from place to place is current, but, on the other hand, diseases have often arrived from elsewhere ... SARS reveals anxieties not only about invasion but proximity in a porous world. The familiar and unknown prove harder to distinguish' (Bush 2003).

The federal SARS report repeats this theme:

> Globalization has made our world smaller as people and goods move more freely and more frequently around the globe. As the world becomes more interconnected, the opportunities for rapid and effective disease spread increase. And as was seen with SARS, travel plays a pivotal role in the rapid dissemination of disease ... Human migration has been a key means for infectious disease transmission throughout recorded history ... The globalization of the food (and feed) trade, while offering many benefits and opportunities, also presents new risks. Because food production, manufacturing, and marketing are now global, infectious agents can be disseminated from the original point of processing and packaging to locations thousands of miles away. (National Advisory Committee on SARS 2003, 16–17)

In a discussion of 'outbreak narratives,' one reporter makes the point that they tend to be 'about fears and anxieties that are associated with globalization. The West has certain fantasies about certain places such as Africa and Asia. The fact that SARS started in Asia has triggered certain responses. Some of the particular concern about SARS is exacerbated by people's fear of certain places that represent the unknown' (Black 2003). Warnings were sounded about the ease of air travel to all parts of the globe, combined with population pressures in parts of Africa and Asia that lead to more human contact with isolated ecosystems and the disease microbes that live within them (Deonandan 2003). Combined with fears about the sincerity of our own governments in maintaining public health is the longstanding cultural theme of fear of Asia and Africa as sources of the unknown. In the globalizing world, the ancient contrast continues to be drawn between the pristine and secure West and the dangerous and insecure Orient. It is from these latter places that the apocalypse will come and the citizens of Canada must ensure that they are positioned to survive.

This sense of insecurity over globalization and Canada's exposure to Old World regions that are traditionally defined as sources of plague and conflict was the basis for a third commonly occurring theme in the SARS coverage – a sense of shock and anger at seeing Canada's largest city listed alongside the Guangdong region, Beijing, Hong Kong, and China's Shanxi Province as a plague site to be avoided by international travellers. 'Torontonians are justifiably asking why one not-very-virulent viral disease has brought this city into disrepair and disrepute when it seems to be contained everywhere except here, Hong Kong and a few provinces in China (according to the World Health Organization which, let us remember, is not exactly a fly-by-night or alarmist outfit). Maybe its because we are the new China ... It seems at times that, as in China, public leaders here are either in denial, tripping over themselves or misguidedly protecting the public from knowing too much' (Gerstel 2003b). Another report warns that Canada must not neglect public health: 'It is precisely this public-health infrastructure that has been allowed to erode in too many poor countries. In China, poor monitoring as well as a culture of official secrecy is partly responsible for SARS escaping from Guangdong region' (Jha 2003).

Overall, the press framing of the SARS crisis and the federal SARS commission report represented and reproduced a number of fears about the unstable nature of biosecurity in post-9/11 Canada. As a test of Canada's readiness to respond to a bioterror attack, a number of problems were identified in the system's ability to identify the event, contain infected patients, transport people safely, and communicate with other health care professionals and disease experts. As well, governments seemed to be scrambling without an effective emergency plan and were blamed for initiating an overly enthusiastic public health cost-cutting agenda over the past decade. In the course of a fairly 'easy' case, Canadian governments failed the test of biogovernance. Consequently, provincial and federal commissions were established to investigate the SARS crisis and recommend policy directions (National Advisory Committee on SARS and Public Health 2003; SARS Commission 2006).

The result was the establishment of a provincial emergency plan to deal more effectively with disease outbreaks and also establishment of the federal Public Health Agency in 2004. Like the 9/11 terrorist attacks, the SARS crisis also forced Canadians to re-evaluate their place in the larger global community. Smug myths of isolation from the problematic areas of the Old World came into question, and once again Canadians were faced with stark evidence that they are part of a global flow of travellers, immigrants, business people, and microbes. Canada, like all other nations, has leaky borders.

What seemed to gall Canadians the most, however, was the stigmatization of having a disease outbreak in their largest city and being unable to contain it immediately. This perceived failure indicated a larger failure to maintain the standards of biosecurity that were being set by other countries such as the United States with its vigorous epidemic control programs and by WHO with its global public health agenda.[3] Canada was compared to a third world nation – a severe symbolic blow to its status in the twenty-first-century global order. In addition to the economic impact on Toronto, Canada's status as a responsible element of the Western security empire came into question. It became apparent that within the new global order, disease is not simply a microorganism or a bodily affliction, but also a marker of a nation's location within the global economic and security hierarchy.

In May 2003, as the SARS epidemic continued to spread in Toronto, a further blow to Canada's place in the biosecurity hierarchy occurred when an Alberta cow was discovered to have bovine spongiform encephalopathy (BSE), popularly known as 'mad cow disease,' which can cause a neuro-degenerative disease in humans called Creutzfeldt-Jakob disease. The Canada/U.S. border was immediately closed to Canadian beef and remained that way until July 2005. During this period, the Canadian beef industry, which sold more than 70 per cent of its cattle to the U.S., lost billions of dollars. In December of the same year, a second Alberta case was found and, just as the border was to be reopened, a third Alberta case was found in January 2005. The ban was lifted in July of that year shortly after an American cow was also found to have BSE (Associated Press 2005). Combined with the SARS outbreak, this further enhanced Canada's global reputation as a disease 'hot zone.'

Canadians, like all other populations within the global security empire, do not want to feel as though they are not part of the elect who could survive the promised epidemics of the future. This is a symbolic outcome of the SARS outbreak and the BSE incidents. Given the ongoing threat of bioterrorism and epidemics, governments must demonstrate to other states and to their citizens that they are responsible by developing and maintaining biosecurity policies and techniques. They must attempt to manage their leaky boundaries to confirm that they can manage the risks of bioterrorism and disease. This is the main reason why the Public Health Agency of Canada was established – it is Canada's attempt to fall in line with international expectations about biosecurity, just as the Department of Public Safety is Canada's response to the global requirement that defence against terrorism be centralized within each individual state and meet the expectations of the emerging international system. Developing these security regimes is a part of the responsibilization of the global security order – it is the price of admission.

Leakiness and the Social Science Fiction of Biosecurity

Despite the implementation of a seemingly impressive biosecurity regime at the national and international levels, the scenarios represented in biothrillers continue to intrude upon any potential sense of satisfaction that a state of security has been achieved. In contemporary fictional scenarios, the biothriller increasingly gives way to biohorror. The quiet, detached, clinical investigation of the film *The Andromeda Strain* in the early 1970s was replaced by the social breakdown of *28 Days Later* (2003) and its sequel *28 Weeks Later* (2007), in which a global epidemic transforms people into murderous, zombie-like creatures. The uninfected must attempt to evade the infected but must also attempt to evade and survive the militarized emergency response forces supposedly sent out to restore order. A note of cynicism about the biosecurity regime is entering public discourse.

There may be some cause for cynicism because, upon closer inspection, the emerging biosecurity system is not designed simply to counter bioterrorism and epidemics; it has a much broader effect – the management of bioagency. Governments throughout the security empire are attempting to seal many types of leaks that occur in their boundaries; this is why Public Safety Canada, Homeland Security in the United States, and similar organizations around the world bring together crime management, public safety, immigration, and border security under one management. Nevertheless, bioagents are very difficult to subject to regimes of biogovernance. Although the Toronto hospitals might have been able to react more quickly to the outbreak of SARS, the biosecurity system of the time could not have prevented SARS – an invisible, quickly reproducing, contagious bioagent – from entering the country. Microbes do not respect international boundaries, nor do other types of agents who act as their vectors including travellers, immigrants, terrorists, scientists, and corporations. Like pathogens, these agents remain stubbornly impervious to the responsibilization techniques of biogovernance and, therefore, can be represented as a threat.

At its highest level of abstraction, the post-9/11 global order is characterized by a two-part hierarchical division of the world into a zone of inclusion, a secure zone made up of nations that are developing security regimes to manage their leaky boundaries, and a zone of exclusion made up of nations that are the source of insecurities. Those within the zone of inclusion operate under an imaginary of control in which government health, military, and police security policies, and their corresponding surveillance technologies are believed to bring a measure of security. It is an imaginary of control because it can never be complete. Borders are always porous; bioagents are changeable, mutable, and mobile and cannot always be detected, and the number of threats continues to expand along with the number of experts defining the problem.

A quick perusal of ongoing media coverage demonstrates continuing insecurities over the porous nature of borders. Since 9/11 and the emergence of an enhanced biosecurity system, a generalized anxiety about bioterrorism and epidemics has not receded. Reports such as this are a continual theme in today's press. Old enemies such as tuberculosis, syphilis, and even the Black Death continue to exist in little known corners of the Old World, threatening to break out again (Immen 2002; Sullivan 2003). New diseases continue to arise as potential threats to biosecurity systems, most recently avian or bird flu, which, once again, originated in China, and the swine flu outbreak of 2009 that originated in Mexico.

However, as in fictional representations of bioterror and epidemics, after more than a decade of disease scares a note of scepticism has entered media discourse – a sense that scientists have 'cried wolf' too often:

> Over the past 15 years public health and infectious disease experts have bombarded us with dire warnings about the 'coming plague.' The usual scenario is for a highly infectious and virulent virus to sweep out of the tropics and overwhelm our medical and biological defences. The combination of population growth, environmental change and rapid international travel has, allegedly, made us increasingly vulnerable to microbiological catastrophe. There has been a long list of candidates, almost a disease a year. However, Marburg, hanta and Ebola viruses, Lassa fever, SARS and the

others have failed to live up to their advance billing. The latest nominee is influenza, a disease we thought we knew from the annual winter flu season. But no, a new superflu will cause a super epidemic, a pandemic, that will kill many millions, maybe even billions, and bring our economies to a standstill and modern society to its knees. (Schabas 2005)

Like bioterrorism, the threat of epidemics is social science fiction. We are uncertain where the boundaries lie between present actuality and future threat and between fact and fiction.

As a result of these uncertainties and the ongoing leakiness of borders, borders themselves are increasingly sites of tension and fear, even within the security empire. For example, immediately after the attack on the World Trade Center, the U.S./Canada border became tightly restricted and patrolled by American border guards. Over time, the restrictions have not eased appreciably and new restrictions continue to appear. In 2003, Michigan senators lobbied the U.S. federal government to require inspections of Toronto garbage trucks carrying garbage to Michigan landfills, stating that the garbage could contain bioterrorism threats (Lu 2003). In September 2006, the U.S. Department of Agriculture announced that in an 'emergency action' it would levy a per-trip surcharge on all air travellers and commercial cargo transporters coming from Canada. Referring to Canada as a potential conduit for bioterrorism, pests, and disease, the money raised from this surcharge would be used to fund an expanded inspection program to screen travellers and commercial shippers for biohazards (Chase 2006). In July 2007, the Canadian government began to arm its border crossing and marine port of entry guards and began hiring 400 new guards to put more security on the border. Tensions are not easing over time and the sense of insecurity grows as more and more possible biosecurity and other security threats are defined.

Given the level of global interconnectedness and the invisible and mobile nature of bioagents, biosecurity and bioterrorism remain social science fictions that inhabit the indeterminate space between present and future, fiction and non-fiction. This space of uncertainty becomes the ground for a biogovernmental intervention in the form of a new biosecurity system designed to manage future possibilities by institutionalizing certain present practices. The constant threat of dangerous bioagents in the form of terrorists, microbes, and travellers to and from the Orient has been normalized through fictional, media, and policy representation that contribute to an institutionalization of biosecurity. Steps have been taken to objectify these bioagents, and the means to counteract them, through a global outbreak surveillance system and through a discourse of security that enables the increasing bureaucratic centralization of crime management, intelligence services, military force, border patrol, and health monitoring. The power of the security discourse blunts any significant politicization of corresponding issues of individual privacy, bodily integrity, and rule of law. However, the biogovernance of epidemics and bioterrorism is not yet complete. Borders and bodies are still leaky and scientists, governments, travellers, and other bioagents are not thoroughly responsibilized. These will be the ongoing targets of biosecurity in the years to come.

CHAPTER SIX

Conclusion: Becoming Biosubjects

Then we can feel confident your so-called biological crisis is over?
Senator from Vermont in Wise, *The Andromeda Strain* (1971)

The province of Saskatchewan is famous for the strength of its prairie winds, and they must have been blowing hard in 1996. That was the year that Percy Schmeiser, a farmer of more than fifty years, claimed that, unbeknownst to him, genetically modified canola seeds blew onto his land. Agribusiness giant Monsanto Corporation claimed that its canola seed was resistant to the herbicide Roundup, also produced by Monsanto. This allowed farmers who planted what was marketed as Roundup Ready Canola (RRC) to douse their fields in Roundup herbicide, killing all other plants but leaving the canola undamaged. Not surprisingly, Monsanto holds a patent on the genetically modified genes and cells in the plant. In keeping with the legal ruling in the Oncomouse decision discussed in chapter 3, however, they do not own a patent on the canola plants themselves.

In 1996, approximately six hundred Canadian farmers planted RRC on 50,000 acres. By 2000, farmers had planted 4.5 to 5 million acres, constituting nearly 40 per cent of all the canola grown in Canada (none of which would be labelled as genetically modified as it entered into the consumption market in Canada).[1] In the mid-1990s, five farmers in Schmeiser's area began to grow RRC. To do so, they had to license the genetically altered seed from Monsanto and sign a Technology Use Agreement for its use, which, among other things, restricted them from saving any of the seeds for replanting in future years, prevented them from selling their crop to any but a Monsanto-approved purchaser, and gave permission for Monsanto staff to regularly test their fields. Percy Schmeiser never purchased RRC, nor did he obtain a licence from Monsanto to plant it. In 1997, however, Monsanto received a hot tip on one of its 1-800 neighbourhood 'snitch lines' that Schmeiser had RRC in his field. The snitch lines were part of a system put into place by Monsanto to protect its patented investment. Its investigators conducted tests on Schmeiser's field and in the spring of 1998 he was informed by a Monsanto representative that they believed he was in possession of RRC. There are only two ways to recognize RRC – either by genetic testing for the altered gene or by spraying the plant with Roundup herbicide to see if it dies. Schmeiser's fields were tested again in 1998 and 95–98 per cent of his 1,000 acres of canola was RRC plants. Schmeiser insisted throughout that the RRC seed had blown onto his land from his neighbours' fields.

Monsanto holds patent 1,313,830 for glyphosate-resistant plants – those plants with a dramatically increased tolerance to herbicides containing glyphosate. It brought a lawsuit for patent infringement against Percy Schmeiser alleging that he had 'used' their patented invention without authorization. In 2001, the Federal Court upheld the validity of Monsanto's patent and the infringement allegation, ordering Schmeiser to pay damages of approximately $15,000 (*Monsanto Canada Inc. v. Schmeiser* 2001). Schmeiser appealed to the Federal Court of Appeal, and in 2002 that court's unanimous decision rejected all of Schmeiser's arguments (*Monsanto Canada Inc. v. Schmeiser* 2002). Finally, like his biosubject kin, Oncomouse, he went to the Supreme Court of Canada (*Monsanto Canada Inc. v. Schmeiser* 2004). The Supreme Court of Canada in another 5–4 decision held that Schmeiser had used the invention of Monsanto by having the plants in his field. The dissenting judges suggested that the effect of the majority's holding was to give Monsanto a de facto patent on the plant itself (a higher life form), and not just the altered genes and cells. The majority contended that the patent was meaningless if Monsanto could not enforce its rights 'in the field.'

In the meantime, however, this case had become a public relations nightmare for Monsanto. This was another classic David and Goliath narrative with the wily Schmeiser playing the role of the small, local farmer being oppressed by a transnational agribusiness corporation engaged in genetically modifying the food supply to unknown effect in the absence of government regulation. Further, various farmers groups around the world tuned into the case and became involved; Schmeiser even stopped farming for a few years in order to travel around the world on the anti-GMO and farmers' rights speakers circuit. Organizations involved in farmers' rights and opposed to genetically modified foodstuff, both nationally and globally, hopped onto the bandwagon. The vociferous cries for Parliament to legislate once again echoed through the press – the human biosubject fighting life form patents seemingly more interesting than his rodent predecessor.

And in another turn in an already strange tale, while Goliath may have won the legal battle, ironically, on the issue of damages, it was David who came out on top. The Supreme Court held that because Schmeiser had not used Roundup herbicide, he had not actually profited from growing RRC, namely from using the invention. As a result, Monsanto was not entitled to any of Schmeiser's profits from that year and they received no damages for their patent infringement. The final twist is that the whole dispute was ultimately rendered moot by a succeeding genetic invention, fondly dubbed in the media the 'terminator seed.' Monsanto (and others) developed a genetically modified seed that reproduces only once and then is sterile, rendering futile any attempts by farmers to save seeds.

In this example, we see issues from the dramatic spread of genetically modified organisms in our food supply, to the global implications of biotechnological practices, to the incredible stakes of such a biovaluable commodity, to the erosion of traditional practices and communities. However, we suggest that, on the basis of the analysis that we have developed in the preceding chapters, it is both provocative and productive to read Percy Schmeiser, Monsanto Corporation, and Roundup Ready Canola, itself, all as biosubjects caught up in a web of relations structured by many of the social science fictions at work in Canada's engagement with biotechnology.

Social Science Fictions Redux

Monsanto's regime of snitch lines, investigators, genetic testing, and monitoring – all contractually guaranteed – form a massive system of genetic surveillance. In securing the consent of the farmers to their own monitoring, they are responsibilized by Monsanto into becoming elements of the surveillance system, ever vigilant against their neighbours, a rural panopticon. This imbrication relies upon their recognizable economic stake in ensuring that all have to pay for the privilege of farming RRC. Yet, as Monsanto learned to its detriment, such 'old-fashioned' regimes of surveillance cannot contain all the risks of biotechnology. Seeds, and the farmers refusing to act 'responsibly,' escaped the dragnet.

The social science fiction of biotechnological surveillance, as discussed in the case of criminal justice in chapter 2, is the elimination of the gap between the body and the code. This manifests in practical terms in the desire by law enforcement officials to dramatically reduce the temporal lag between the occurrence of crime and the identification of perpetrators through their already-stored codes. If one did not have to deal with the complicated and messy bodies standing in the way of that practice, with their accompanying bevy of legal rights, the detection of criminals would be instantaneous. Further along the social science fiction spectrum, the elimination of the gap between body and code would ideally produce a scenario where criminality is not marked on the body as material infrastructure (or in its mind), but already in genetic code. If criminality can be circumscribed within subjects' DNA, then the ideal is to identify gene sequences more likely to produce criminal behaviour and eliminate or modify them.

In the Schmeiser example, the shift from analogue modes of surveillance requiring human agents, the securing of consent of the participants, and ongoing action by the investigator has been replaced by genetic surveillance, which much more closely approaches the social science fiction of codes replacing body. Terminator seed technology alters the body of the seed at a genetic level to render it non-reproducing and hence non-reproducible. This is a much more seamless mode of surveillance, coding it right into the genetic structure of the being and, therefore, requiring farmers to re-license their access to the seeds every year. There is no longer any need for snitch lines, investigators, and genetic testing. In terminator seed technology, the social science fiction of complete genetic surveillance has been achieved.

The courts in Schmeiser were on the horns of a legal, if not social or moral, dilemma. If they held that Schmeiser had not infringed upon Monsanto's patent, then essentially that reduced Monsanto's patent rights in the Roundup Ready genes to their use only in a laboratory. Only a molecular scientist would be capable of infringing on the patent rights. On the other hand, if they held, as they did, that Schmeiser had 'used' Monsanto's patented genes in his cultivation of RRC, which in any ordinary understanding of the notion he had, then the effect of that is to grant Monsanto control over all existing and possible plants containing the Roundup Ready genes. The social science fiction of the higher life form as property, having been acknowledged but forestalled in the Oncomouse saga, becomes scientific, if not social, fact as a result of the Schmeiser conflict.

While coverage of the Schmeiser case was greater than that of the Oncomouse, there was still limited public or media outcry to the Supreme Court of Canada decision, which resulted in the de facto patenting of a higher life form. Instead, the outcry was focused on the value, from health and economy perspectives, of genetically modified foods. This allowed the case to be figured politically, within a pre-existing and distinctly Canadian media discourse on biotechnology, one that repeatedly sidesteps moral issues about the definition of life in favour of more pragmatic concerns about health (Sullivan 2005). While the Oncomouse had raised the social science fictional spectre of higher life forms as property, canola is merely a plant and the leap to the human seems more remote. Thus it is interesting that the Schmeiser case received far more attention than that of the Oncomouse, as if it were a safer and easier conundrum to solve. The five conditions of possibility for propertizing the biosubject, discussed in chapter 4, have already shaped the public consciousness of Canadians. The Schmeiser case transforms the social science fiction of the higher life form as property into social and scientific fact. All that remains is to close the gap to the seemingly more distant social science fiction of the human being as property.

A key social and economic issue at the heart of the Schmeiser case that played out globally was the changing role of the farmer and his or her relationship with the land, the activities of plant reproduction, and a rural lifestyle. Already under attack from corporatization and so-called factory farming, this case was seen by many as the final nail in the coffin of traditional farming and therefore of a highly romanticized way of life for Canadian citizens that privileged sovereignty of individuals over property. Now those individuals themselves were being responsibilized into the property of huge biotechnological companies. Under the twinned rubric of efficiency and productivity, farmers in Canada over the course of the 1990s began to radically alter how they farmed. Rather than purchasing seed, planting it, saving it, replanting, and, in fact, often engaging in their own hybridizing and thus genetic manipulation, they were now more likely to license their seed from a supplier, be prevented from saving or manipulating it, and then be required to repurchase seed the following year.

Reproduction had been taken out of the hands of the farmer and relocated to the corporate offices of Monsanto, through the vehicle of the licensing agreement and the carrot of the 'improved product.' Apparent, here, is a parallel to the social science fiction of reproduction severed from the woman's body that we discussed in chapter 3. There, again

bolstered by techniques of biotechnologies and contracts, human reproduction is refigured along the lines of efficiency and productivity, removed from the complicating elements of female subjectivity and embodiment. This is a shift from reproduction to replication. As Maureen
McTeer had warned, farmers and women alike are responsibilized into a
system that requires their submission to a biotechnological regime of
expertise. Yet while women can neither profit from nor control their
reproductive material, Monsanto can.

With terminator seeds, this takes the social science fiction to the next
level, where control over reproduction is solely biotechnological and is
therefore complete. The seeds reproduce and cease reproducing on cor-
porate-scientific command. And yet again because the technology involves
seeds rather than babies, the stakes seem less extreme. There have been
no panicked attempts at legislation by the Canadian state as there have
been, repeatedly, in the domain of human reproduction. In the domain
of agriculture, the Canadian government seems content to let the market
and science manage replication. Yet both moves from reproduction to
replication – in seeds and embryos alike – rest on the same social science
fiction of de-natured, disembodied, and scientized reproduction.

Much of the appeal of the Schmeiser case to the general public and
the media was the implausible explanations that Percy Schmeiser offered
for the RRC on his land. Experts in court testified that it was scientifically
impossible for that percentage of his crop to be RRC merely from the
agency of the seeds and wind themselves. Human agency was required;
yet Schmeiser refused to admit his role. What the scientific experts were
forced to acknowledge, however (as they had had to in relation to complaints by organic farmers made previously), was that seeds do transgress
the boundaries of property lines. The leaky borders within farming communities were revealed in the mobility, one might say bioagency, of genetically modified canola seeds. Thus RRC also participates in the social
science fiction of the contamination resulting from the combination of
leaky borders and genetically modified beings/pathogens.

As discussed in chapter 5, microscopic bioagents are notoriously difficult to bring within effective regimes of biogovernance. They are invisible
to the naked eye, they are available to the public, they can exist en masse,
and once their reproduction is released into the environment, it is not
subject to human control. Unpredictable environmental elements from
the wind, to terrorists, to the migration of people all render the genetic
pathogen essentially ungovernable. And of course, when this happens on
a global level, the challenges are compounded, seemingly exponentially.

The deep-seated cultural fears both accessed and rewritten in the social science fiction of global devastation is visible in the case of genetically modified foodstuffs in the panic-mongering discourse of 'Frankenfood,' which serves as a dominant critical frame. From the off-putting image of fish genes in the FlavrSavr tomato to the urban legends surrounding children dying from peanut allergies while unsuspectingly eating genetically modified apples, the invisibility of the threat (both in the products themselves and in the labelling), as well as the lack of certainty and longitudinal scientific study, make ripe the conditions for public concern and even panic. This gap between present knowledge and unknown future outcome, between present practice and future risk, reproduces the social science fiction of global pandemic.

Not incidentally, Percy Schmeiser was able to fund the costs of his legal appeals through becoming an internationally renowned advocate for farmers' rights in opposition to global agribusiness. Just as genetic pathogens will not respect national borders or social distinctions between first and third worlds, neither do plants or, for that matter, transnational corporations. In this way, Schmeiser could be figured as advocating outmoded methods of farming more appropriate to a third world than first world. The social science fiction reinforces those socio-economic distinctions, but the wind blows both ways. Just as immigrants to Canada, particularly from the Pacific Rim, were cast as a threat and a source of plague to the nation during the SARS crisis, Monsanto and other agribusiness behemoths are cast as a source of a parallel genetically modified scourge by a growing global coalition of activists acting within the developing world. The recently successful challenge to W.R. Grace's patent on the Indian neem plant by Vandana Shiva (2005) and other activists supported by farmers' rights groups and green parties and organizations around the world, is evidence of such a frame. Cast as biopiracy, the actions of first world pharmaceutical and other corporations are increasingly being constrained by local governments and global public opinion. At the same time, third world populations and their accompanying germs continue to be viewed with fear in the first world.

The Schmeiser versus Monsanto example demonstrates the continuing power of the social science fictions identified in each of our case studies. From seamless genetic surveillance to the higher life form as property to reproduction severed completely from the body to genetically induced pestilence, these social science fictions do the conceptual work of bridging present social and scientific fact with future possibility and risk. They mystify the power of biotechnology and its holders and,

whether figured as utopian or dystopian, posit a radical social reconfig-
uration. We are not suggesting that these social science fictions are
specific to Canada's attempt to interpret biotechnologies; however, we
do posit that they play themselves out in different ways in specific national contexts. Most importantly, however, in order to reflect critically
on the operation of any biotechnology, its implementation, its stakes,
and its public, popular, and institutional reception, one cannot ignore
the cultural power of these archetypical narratives.

Contingent Body Ontologies

In chapter 1 we mapped a series of changes that we suggest lay the ground-
work for a new way of thinking about subjectivity through the molecular
optic. First, we suggested that the material substrate of the subject, the
body, has undergone a shift in its underlying ontology, from bodily integrity to contingency. Contingent bodies are divisible, namely separable
from any larger metaphysical whole; they are permeable, no longer
bounded by the physical or legal skin; and finally, they are volatile and
mobile, capable of appearing in unlikely places and then reappearing
elsewhere. They are bodies fundamentally altered at a molecular level
through the practices of capitalist technoscience. The contingent body
as a ground of subjectivity has therefore clearly been severed from the
human being.

In the Schmeiser case, the contingent body of the Roundup Ready
Canola can be recognized only in and through its codification; it has an
unpredictable motility and its identity has been severed from its capacity
for auto-production. In the case of DNA databanking and criminal justice, through the traces of the material body that has passed through the
crime scene, the contingent body of the offender can be reconstructed
in the lab, on-site, or in the courtroom. The bodies of all Canadian citizens have been rendered more contingent in their permeability – required to submit to the breaching of their physicality and privacy in order to offer up their genetic material for codification and comparison.

The Oncomouse and embryo are contingent bodies par excellence.
Simultaneously legal subjects and scientific resources, potential beings
and medical objects, their bodies have been divided and de-reified.
Their close relationship with the adult human being (as higher life form
kin and as reproductive possibility, respectively) means that their very
existence troubles the bounded body of the liberal, individualist subject.
Both bear rights defended by advocacy groups to which they cannot

belong, and yet both are unproblematically, simultaneously recognized by the public, governments, and the law as property. They demonstrate clearly the hybridity of the contingent body.

Finally, the messiness of the contingent body cannot be contained within either the symbolic or material realms of bioviral agents, intriguingly, both at the individual and national levels. They are mobile and volatile in the extreme, regularly exceeding the spaces of the laboratory and the nation. Global tourism, immigration, and business produce the conditions for travelling humans to serve as the vectors for the contingent bodies of rapidly mutating diseases. They can lie dormant for years and then suddenly appear without warning. They can also evolve rapidly and in unpredictable ways, furthering their claim to contingency. The nation, always itself a contingent entity, particularly it seems in Canada, remains permeable, with biogovernmental action always, by definition, incomplete.

Governing the Ungovernable?

We mobilize the concept of biogovernance as a means of articulating the relationships between varying sets of social actors involved in different sites of biotechnology and the Canadian state. The case studies reveal a wide range of biogovernmental tactics and an even wider array of biogovernmental effects. Below, we review the components of biogovernmental practice in relation to the Schmeiser example and our case studies, arguing that biogovernance is an always ongoing and negotiated process among governmental, legal, and individual actions.

The traces of biogovernmental tactics are revealed in Monsanto's responsibilization of Canadian farmers to watch their neighbours and participate in Monsanto's surveillance system under the auspices of protecting their own economic interest. This is very similar to the ways in which non-accused citizens are implicated in DNA dragnets, not only by police, but by their fellow community members. Obviously this has a detrimental impact upon community solidarity, forcing people to act in their own self-interest and in the interest of a greater abstraction, be that economics or crime management. As we point out in chapter 3, women have been increasingly responsibilized in their obligations to manage their pregnancies for the health of the foetus, as well as the health of the family and nation. Pregnancy, constructed as a risky practice, is now seen as something in which both the medical profession and state have a vested interest in protecting. The analysis of biosecurity in chapter 5 reveals that when international biogovernance fails, individual states must take up the responsibility for their leaky borders. Further, as the SARS crisis showed, bioagents inevitably slip through and at that point individual citizens bear much of the responsibility to be healthy, to stay out of diseased zones, and to be ever vigilant about the sources of disease around them (including fellow citizens).

The Schmeiser example also reveals the extent of the shift to privatization in that the most significant developments in the bioagriculture domain are taking place in and through the private sector, a sector that turns to intellectual property and contract laws for its legitimation and enforcement. The private sector has also been very resistant to government regulation of its activities, as is apparent in the ongoing Canadian disputes around whether or not genetically modified foodstuffs should contain warning labels (Andrée 2002). Privatization is arguably responsible for the opening up of abortion rights in Canada, but, at the same time, raises the spectre of attributes-on-demand-style fertility clinics à la *Gattaca*. The Oncomouse case seems to fulfil the technologically determinist claim that science (always private sector science) keeps marching forward, despite governments, social movements, or the public, and it is only a matter of time until someone clones a human being. Private sector science is thus figured as a fundamental force for good, but bringing with it some inadvertently negative consequences, regularly exaggerated by B-movies and genre fiction.

Within our biogovernmental analysis, both politicization and normalization involve the management of symbolic frames to contain the risks of dissent. This process takes place in Canada in the media, the courts, policy mechanisms, and legislatures. Percy Schmeiser became active in managing his own image and issues, although he had to contend with a press that did not always accept his self-characterization. Monsanto was quickly forced to realize that this was not merely a patent dispute, but a dispute for the minds and hearts of Canadians. It had to attempt to shape the biotechnological imaginary. Apparent throughout our case studies are the fundamental challenges of both politicization and normalization: there are two fundamentally opposed ontologies at work. There is no shared language that can resolve their differences; there are no political techniques that can produce consensus. In both criminal justice and reproductive issues there is a fundamental dispute between bodily integrity and contingent bodies, between citizens and biosubjects. In the case of the Oncomouse, the continued playing out (and reification) of the distinction between nature and culture is devolving and yet remains the ongoing axis of debate. On the issue of bioterrorism, two different models of biosecurity have been proposed. The UN and WHO model is one of global *community health* that seeks security *of* the global body by responsibilizing states to sign treaties that ban biological weapons, regulate scientists, and share public health resources in a global enterprise. The other, more successful model has been a *disease* model designed to produce security *against* foreign bodies. Consequently, ever-tighter security regimes are normalized within many countries while the power of the security discourse ensures an absence of politicization within fearful populations.

The objectification element of biogovernance is a precursor to what we suggest is a broader epistemological shift towards visibility. The entire validity of biotechnological practices – be they legal, scientific, or contractual – turns on the valorization and valuation of the capacity of molecular science to render what has been historically invisible, visible. In both Monsanto's and Harvard's cases, this capacity for rendering, for representing, results directly in property rights and economic value. Monsanto and Harvard are read as the authors of their creatures. In criminal justice, visioning affects conviction rates in significant ways, and the forms of expertise legitimated to see and thus speak. Embryology made the human embryo visible in mid-1800s, but with the aid of genetic science, being able to 'see' the embryo at a molecular level permits its construction by science for the first time. Finally, the visibility of the pathogen is twofold and contradictory: its creation is enabled by molecular scientific techniques but its threat is largely constituted in its continued invisibility.

Given the mixed success of biogovernance in our case studies, we come back to the issue with which we began this section: is biotechnology governable? Certainly there are factors that contribute to the tensions and instability in any attempt by the Canadian government, or any other state, to manage the risks posed by biotechnologies. These include the element of privatization, the limits of scientific reflexivity, the global nature of the biotechnology industry, and biotechnology's instability as an object of governance. The private sector offers the expertise and financial resources driving the biotechnology industry. As with the Human Genome Project, the public sector inevitably lags behind. When the interests of the private sector and governments coincide, as they seemed to do in DNA databanking, for example, then governance is more effective. When those interests do not mesh, as in the ideological division between the Supreme Court of Canada and the Canadian biotechnology industry on patents of higher life forms, effective intervention by the state is more difficult.

While many scholars have claimed that society is now in a moment of high reflexivity towards science and technology, we suggest that that moment has passed, or perhaps never really occurred, with respect to biotechnology. While there was a global crisis of questioning biotechnology in the wake of Dolly, arguably the scientific community has been effective in both discrediting and waiting out that panic. Society is definitely now in a post-Dolly era where biotechnology has retreated behind the curtain again, has successfully framed itself in relation to public health, and has effectively contained its monopoly on expertise. The black box has closed once more.

As a number of the case studies and the Schmeiser example suggest, biotechnology exceeds national governance. It is a global issue and yet the international efforts to address biotechnology have been almost exclusively in support of its further development and the protection of the private sector. Resistance to the global effects of biotechnological developments has had to come from coalitions of social movements, grassroots organizations, and alternative political communities, with uneven success. The effectiveness of any one nation's regulatory actions can be undercut by international pressure or by the relocation of the activity in question to a more sympathetic national jurisdiction. As well, the largely successful attempts by the biotechnology industry to align their interests with the economic health of the nation (with the important exception of the Oncomouse case) has meant that movements towards trade liberalization are simultaneously also moves toward biotechnological liberalization.

Lastly, we argue most importantly, the very nature of the biotechnological imaginary renders biotechnology a problematic object of governance. The object of governance is unclear because it is a hybrid of future possibilities and present actualities. The Canadian government faced difficulties in bringing in legislation addressing human cloning, chimeras, and in vitro fertilization, in part because a number of those practices could be discredited as neither yet happening nor even likely to happen. On the other hand, the Canadian and other governments were unable to respond to disease threats effectively. They seem caught off-guard, but that is difficult to comprehend, given that they have been a part of our biotechnological imaginary for years through popular culture. In contrast, the CSI effect poses a challenge to governing agencies to try to 'catch up' to the fictional biotechnological imaginary in their governmental practices. Governments, one way or another, are having continually to deal with the gap between present and future, between science fiction and science fact, that is at the heart of biotechnology as an object of governance. And this is a challenge that the Canadian government has responded to in varying ways, depending upon the political stakes and historical context of the particular social site. So, more important than our initial question of whether biotechnology is governable or not, is the question that this book has taken up: *how* is biotechnology governed?

Molecular Optics, Molecular Knowing

In the ongoing court battles of Percy Schmeiser, it is remarkable that, throughout, he is figured as an expert in husbandry but not as an expert in growing canola. Instead, agricultural and molecular expertise is called by both sides at the trial in order to make claims about the growth patterns of canola, the trajectory of spread of the genetically modified seeds, the impacts of herbicides, and so on. A focus on the molecular level necessarily privileges certain forms of expertise that bring with them particular ways of seeing and thus knowing. The case becomes centred upon genetics rather than farming practices, and therefore Schmeiser's experience is only that – experience, not expertise. His testimony is treated as a factual account of his activities and he is excluded from offering interpretation and evaluation; those modes of knowledge-making are reserved for the experts recognized by the courts.

This very much echoes the presentation of DNA evidence in criminal trials. Eye-witness accounts have been dramatically weakened in their probative value in the face of the charismatic authority of DNA evidence. Increasingly forensic scientists are providing the ideal optic for courtroom evidence and there is a real fear that scientific expertise is supplanting legal expertise as the basis for trial decisions. Why does one need the adversarial process, lawyers, judges, and precedent, if one has convincing and almost error-free forensic evidence?

Molecular visibility is a necessary precondition of the higher life form as invention. At a practical level, mice, humans, and plants are found in nature. This would typically have located them outside the purview of intellectual property claims. However, the labour of the molecular scientist, which is fundamentally and often exclusively the labour of 'making visible,' transforms nature into invention. Through the romantic myth of authorship, this labour of visioning is taken as the political and legal basis for an act of creation. In this way, a new category of expert becomes inventor and author. A new object of knowledge emerges, an invention, writing over the higher life form as it previously existed.

In the abortion debates in Canada, we can see that women's rights to bodily integrity and dignity were ultimately the basis for overturning the criminalization of abortion. At some level, we suggest, this recognizes the expertise of women over their bodies, including the fetuses within them. Reproduction was about pregnant women. However, over the past twenty years there has been a shift in the object of knowledge, from pregnancy to embryo; in the techniques of knowledge, from material, embodied practice to mediating the visibility of the hidden body; and in the holders of knowledge, from women and doctors to the medical profession, embryology as a discipline, and genetic scientists. Increasingly, women are figured not as experts of fertility, infertility, and pregnancy, but as a resource to be deployed by a battery of scientific experts in the achievement of family, and thus national, health.

In the case of disease microbes, the pathogens are visible in a laboratory, but while travelling along the necessary vectors seem to disappear. Therefore, the invisibility of pathogens as bioagents poses a crisis. Frontline health-care workers are not molecular scientists, and yet to be effective, they need to be able to recognize and identify the pathogens. Microbe detection technologies simply collect air samples that must be brought back to the laboratory for analysis. This presents an inevitable window in time between transmission, detection, and action that grounds much of the fear and tension in the popular culture of biotechnological thrillers. Experts are forced to focus their vision on the vectors, the risky humans, rather than the active agents of the diseases themselves.

Portrait of the Biosubject

The protagonist or even anti-hero in the Schmeiser versus Monsanto tale is the Roundup Ready Canola itself. It travels on its own (although not likely with the efficiency Schmeiser claimed for it); it contains within it the capacity to resist poisons that kill all other plants; it is disguised in the appearance of a non-modified canola plant; and it is the agent provocateur of a bitter legal and political dispute between one way of life and another. In short, it is a biosubject.

The criminal is an obvious biosubject when one considers DNA databanking, but less obvious and arguably more significant (in part because of the lack of notice it has received) is the interpellation of the everyday citizen as biosubject of criminal justice. Citizens are now always potentially virtual, in that they are always expected to be willing to make themselves and their bodies available for inspection and codification by the state. This is because criminality is perceived by the state as a kind of social pathogen, always lurking, unseen, in the social body. At the same time, the broader biotechnological imaginary suggests that criminality potentially lurks in any or all of us through our DNA.

Joining canola as a biosubject is, of course, the Oncomouse. Simultaneously, a higher life form, a mammal, and a being capable of being defended by animal rights activists, being a pet, and contracting life-threatening illnesses, the cancerous mouse is also an object, invention, property, and resource in science's ongoing battle against disease in humans. Its very hybrid nature made it a most effective agent in catalyzing the patent disputes, much more effective, for example, than the bacteria of previous contests. Genetically modified before birth, it carries with it the invisible stamp of scientific intervention, and that status lasts through generations. Our own status as animals and our traditional superiority to other animals is in jeopardy if we cling to the higher and lower life form distinction because all higher life forms are biojuridical subjects.

Similarly, and even further along the continuum, the embryo is both potential human being and medical science fodder. Its agency is located both in its own potential to develop 'naturally' into something else and also in the way it is mobilized by political groups. Alternately figured as person and as waste, the embryo is a powerful actor in social debates from abortion, to surrogacy, to fertility, to cloning. Originally bound to the body of another, it is now mobile, able to exist in a variety of natural and artificial settings. Finally, its homologous relationship with the human being produces much of its volatility.

Our final example of biosubjectivity, the pathogen, is a highly mobile, virile, uncontrollable biosubject. Its agency is constituted in its capacity to move along a variety of natural and constructed pathways and in its capacity to cause reactions in others. Entire societies have had to be reordered to accommodate its existence. Even more so than the human embryo, the genetically modified pathogen is a dual creature of both fiction and fact. Like the Oncomouse, it embodies in itself the strongest fears about the uncontrollability of nature and of science run amok. It respects no borders and recognizes no status in its interaction with humans as differentially positioned biosubjects.

So if canola seeds, citizens, mice, human embryos, and pathogens can all be biosubjects along with citizens, what does the biosubject look like? The biosubject is first and foremost an agent in a set of biotechnologized social relations, either able to take action itself or to provoke others to take action on its behalf. It is also always an object, capable of being read as a thing, as property, as the outcome of the actions of others. It is not necessarily a human being or even a legal person. It has been divided, visioned, and represented in biotechnological terms; it is code.

The resulting being is already both subject and object, person and property, nature and invention. The biosubject emerges when we have the capacity to make life, not merely alter it or fuse it with technology. In challenging most if not all of the fundamental dualisms of Western society and the comfortable hierarchies of being in which people have lived, the biosubject causes social anxiety. Biosubjects are, therefore, often at the heart of some of the most intense social, legal, and political debates of this time. Biosubjects are infinitely mutable, replicable, and uncontrollable. They reveal simultaneously the incredible power of 'nature' as well as of 'science.' They benefit from the mystifying power of both biotechnology and nature. They escape the boundaries placed around them. – Petri dish, governmental regulation, Judeo-Christian ethics.

They are also fundamentally unruly subjects. They disrupt social categories, hierarchies, and relations. They demand a response: ethical, social, political. Once they have been spoken, there is no going back; the social terrain changes immediately upon their recognition. They exceed regulation at every turn and yet are not politically predictable. They simultaneously inhabit present and future, rendering the past irrelevant to their interpretation and attempts to control them difficult at best, doomed to failure at worst. They demand, not resistance in any simple or causal sense, but rather contain within them the elements both to bolster and disrupt the power dynamics from which they are produced. The biosubject as cultural figure and as social actor requires new ways of looking to be recognized, altered epistemologies through which to be known, and a reinvented politics through which to be critically evaluated and practised. Whether we like it or not, we are all biosubjects.

Notes

Chapter 1

1 For more on the emergence of social science fiction as a historical and epistemological mode, see Gerlach and Hamilton (2003). The complete special issue *Science Fiction Studies* 30, no. 2, takes up the theme of social science fiction.

2 The work of Edna F. Einsiedel and Linda Goldenberg is an exception, as they have worked to place Canada's experience in the international context. For example, see Einsiedel and Goldberg (2004) and Einsiedel and Timmermans (2005).

3 See, for example, R. Sullivan (2005).

4 Other work focusing on the meaning of the gene examines the troubled status of nature, the changing nature of the body, and the overall cultural frames used to make meanings of biotechnology (e.g., Graham 2002; Hartouni 1997; Marchessault and Sawchuk 2000; Nelkin and Lindee 1995; Robertson et al. 1996; Roof 2007; Turney 1998), while others attempt to theorize a broader philosophical shift in thinking activated by, and manifest in, biotechnology (Thacker 2005).

5 Similarly, Janine Marchessault (2000) demonstrates the challenge that the development of thermodynamics presents to divisions between nature, technology, and the body by linking them to an unseen energy or life force that is bigger than all of them, but requires harnessing for humans to reach their fullest potential (57).

6 See the interesting work of Lily E. Kay (2000) about the historical foundations of that shift.

7 See Gerlach and Hamilton (2005) for some of the broader challenges in governing biotechnology.

8 For more on this process, see the final report of the Canadian Public Health
Association (2001).

Chapter 2

1 The first Canadian case involving DNA testing was the case of *R. v.
Bourguignon* in 1991.
2 In *R. v. Borden*, the accused had been identified and arrested as the perpetrator of a sexual assault in 1989. The police asked him to provide a DNA
sample for that offence, which he agreed to do. DNA from the sample
matched that from a semen sample found at the crime scene. However,
the police also suspected Borden of a previous sexual assault on an elderly
woman at a nursing home. She had been unable to identify her attacker
because he had held a pillow over her face. Police forensics experts compared the DNA from Borden's sample with that taken from a semen sample
at that crime scene as well. It also matched and Borden was charged with
both sexual assaults. At trial he was convicted of both crimes; however, upon
appeal, the court overturned the conviction for the earlier assault. Borden
had not been asked to provide a sample for that crime. In 1994, the Supreme
Court upheld the appeal court's decision. Borden had volunteered a blood
sample relating to a specific charge of sexual assault; the police were
obligated to inform the accused that they intended to use the sample to
investigate a previous crime. There was no consent to this and the evidence
was inadmissible. Borden's acquittal for the first assault was upheld and
Borden was sentenced to only four years in prison for the second assault.
3 Examples of police dramas focusing on the all-too-human elements of
detectives and officers include programs such as *Hill Street Blues* in the 1980s,
NYPD Blue in the 1990s, and *The Shield* in the 2000s – three of the most
popular crime dramas of the last three decades.
4 Sent to guard an exhibit of Canadian gold at the 1904 St Louis World Fair,
Edward Foster attended a talk by Detective John Ferrier of Scotland Yard, a
fingerprint expert. Foster studied under Ferrier and, upon returning to
Canada, found that there was only ambivalent interest among politicians,
judges, and lawyers. With the support of the commissioner of police, Sir
Percy Sherwood, Foster convinced the government to pass an Order in
Council in 1908, allowing for the use of fingerprinting under the *Identification
of Criminals Act* of 1898. Little was done to support the new technology,
however, until 1910 when a fugitive who had killed a constable six years
earlier was finally captured. The public was outraged that he had never been

fingerprinted or photographed and the minister of justice hastily established the Canadian Criminal Identification Bureau with Edward Foster as its head.

5 The first of these was the case of Wilbert Coffin, a Quebec prospector convicted in 1953 of the murder of three American hunters. He loudly proclaimed his innocence, leading the Cabinet to pose a hypothetical question to the Supreme Court: given that Coffin had been denied leave to appeal to the Supreme Court, how would the court have decided if it had heard such an appeal? The court responded that it would have had no reason to overturn the conviction. Coffin was hanged in 1956. The second case was that of Steven Truscott. In 1959, the fourteen-year-old Truscott was sentenced to hang for the murder of a twelve-year-old girl. In 1967, the Supreme Court re-examined the trial evidence and new submissions. The trial verdict was not overturned, but Truscott was paroled two years later.

6 Rubin 'Hurricane' Carter was twice convicted of three murders in Paterson, NJ, in 1966. With the assistance of members from a Canadian commune, two convictions were set aside and the state declined to re-prosecute a third time. He now lives in Toronto. Donald Marshall spent eleven years in prison for the murder of Sandy Seale. Roy Ebsary, the target of a botched robbery attempt by Marshall and Seale on the night of the murder, later confessed to the crime. Steven Truscott was initially sentenced to death for the murder of a classmate in 1959, but the sentence was commuted to life imprisonment. He was acquitted nearly fifty years later, in 2007. Thomas Sophonow was tried three times for the 1981 murder of Barbara Stoppel, eventually being acquitted by the Manitoba Court of Appeal in 1985. In 2000, Winnipeg police announced that new DNA evidence proved his innocence. Wilson Nepoose was convicted of murder in 1987 but a new trial was ordered five years later when a Crown witness admitted to lying on the stand. Nepoose died before the case was re-tried. Susan Nelles, a nurse at the Hospital for Sick Children in Toronto, was charged with the murder of four babies. Police later admitted that they arrested her in part because she requested legal counsel, which they took as an admission of guilt. The charges were thrown out at the preliminary inquiry and no one else has been charged. Richard Norris was sentenced to twenty-three months in prison for sexual assault in 1980. Eleven years later he was cleared after a friend confessed to the crime.

7 Under section 254 of the Criminal Code, a blood sample can be demanded if the offence has been committed within the last two hours, if there is cause to believe that the suspect cannot provide a breath sample, and if there is a qualified medical practitioner present who is satisfied that the suspect will not be harmed.

Chapter 3

1 As McLaren and McLaren note, '[IUDs], because they were not drugs, were at first neither sufficiently investigated or regulated. The distribution of four million Dalkon Shield IUD's between 1971 and 1975, which resulted in a worldwide outbreak of pelvic infections, miscarriages, congenital birth defects, and maternal deaths, cast a shadow over all high-tech contraceptives. The distribution in Canada of the Dalkon Shield IUD ceased in 1974, with the manufacturer in the 1980's placing advertisements in the press calling for all women still using the device to have it removed at all costs' (1986, 140).
2 More clinics followed in Winnipeg and Toronto in 1983, Halifax in 1989, and Fredericton in 1994 (Morton 1992).
3 See the interesting discussion of this in Morton (1992).
4 See discussion by Eileen V. Fegan (1996).
5 Daigle was hailed by pro-choice advocates as doing what was necessary in defying the injunction, whereas anti-abortion activists called her a murderer and a law-breaker. But while Daigle's image may have been tarnished by her actions, Tremblay was also a problematic figure for the anti-abortion camp. He made frequent impolitic comments in the media during the trial, and later threatened Daigle on the steps of the Supreme Court. 'Chantale did not kill my child, but our child, and for that I will never forgive her. It is a murder she committed, and she will pay one day' (in Wallace and van Dusen 1989, 13).
6 See interesting discussions of the commission and its work in Basen, Eichler, and Lippman (1994). In particular, see the chapters by Magrit Eichler, Anonymous, Christine Massey, and Louise Vandelac.

Chapter 4

1 Mickey Mouse came to be associated with the recent extension of the term of copyright in the United States. While it was officially called the *Sonny Bono Copyright Extension Act* of 1998, it had been lobbied for very extensively by Walt Disney Co. because it would otherwise have lost copyright protection of some vintage footage of Mickey Mouse. The amendment extends the copyright term from life of the author plus fifty years to life plus seventy years.
2 In early 1993, the Department of Consumer and Corporate Affairs appointed senior policy analyst Brian Botting to solicit the views of interest groups and the public on the Patent Office's policy of refusing plant and animal patents. Three years after the study was begun, then under the auspices of Industry

Canada, no report was produced. Our research suggests that the study was never completed.

3 In the United States, courts have argued this would be illegal, on the basis of the Thirteenth Amendment to the Constitution prohibiting slavery. In Canada, there is disagreement whether section 7 of the *Charter of Rights and Freedoms*, guaranteeing the right to liberty, could serve as the basis for a prohibition on the patenting of a human being. In fact, the majority of the Supreme Court suggested that in the event that higher life forms are found to be patentable subject matter, there is presently no legal basis upon which to distinguish human beings from other life forms. Human beings in Canada, therefore, could be patentable.

4 A press search of the period reveals only four pieces that circulated across a variety of different papers around the country.

5 Studies focused on a variety of themes, including a history of the patent system in Canada, international comparisons, patenting genes, the use of animals in research, economic arguments, agricultural issues, human biological materials, ethical issues, competition law issues, and human rights issues.

The consultations included a CEO/President Briefing in Ottawa on 29 September 2000 with sixteen participants, involving primarily representatives from the pharmaceutical, bio-pharmaceutical, and plant and animal biotechnologies industries; a non-governmental organization hearing on 23 November 2000 with seventeen participants; and a scientific researcher electronic forum in February 2001 with thirteen participants.

6 This propertization of the self is arguably a logical next step in the shifts in subjectivity already present in advanced modern societies, with their elements of enterprise, self-actualization, prudence, and responsibility (Novas and Rose 2000; Rose 2001).

Chapter 5

1 These target areas for counterterrorist action include denying financial support by monitoring and more strictly regulating the transfer of funds through legitimate and illegitimate forms of money exchange from private business and charities to global drug networks; regulating more strictly the sale of conventional weapons to other countries and to non-state actors; reducing recruitment opportunities by monitoring terrorist use of the Internet and criminalizing the use of hate propaganda; denying terrorists access to travel by paying closer attention to individual travel bans and by improving detection of identity theft and fraudulent travel documents;

deterring states from supporting terrorist groups through sanctions; and denying access to nuclear, chemical, and biological weapons of mass destruction.

2 As a measure of how vigorously the U.S. government is preparing for biological warfare defence, prior to 9/11, its stockpile of smallpox vaccine was only 15 million doses.

3 The need to better conform to emerging international standards set by the United States and the World Health Organization was recognized in the National Advisory Committee on SARS report:

> Working collaboratively with international bodies is also a key component to dealing effectively with infectious diseases. Canada is in regular contact with the World Health Organization and the US Centers for Disease Control and Prevention in its day-to-day business of conducting disease surveillance ... Like many other countries, the USA is in the process of improving its national capacities for disease surveillance, prevention, and control. It has developed a strategic plan for preventing emerging infectious diseases, the pillars of which are surveillance and response, applied research, infrastructure and training, and prevention and control. The CDC seeks to improve epidemiologic capacity, surge capacity, communications, and the supply of appropriate and adequate equipment and training ... The experience of the SARS outbreak has renewed calls for a Canadian version of the US Centers for Disease Control and Prevention to improve coordination of public health across Canada. (National Advisory Committee on SARS and Public Health 2003, 27–8)

Chapter 6

1 See the excellent, comprehensive article on biopower and the regulation of genetically modified foodstuffs in Canada by Andrée (2002).

References

Abraham, C. 2002. The lab rat pack. *Globe and Mail*, 31 August.

Abraham, C., and L. Priest. 2003. Cutbacks fed SARS calamity, critics say. *Globe and Mail*, 3 May.

Agamben, G. 2000. *Means without end: Notes on politics*. Minneapolis: University of Minnesota Press.

Agar, N. 2004. *Liberal eugenics: In defence of human enhancement*. Oxford: Blackwell Publishing.

Alibek, K., with S. Handelman. 1999. *Biohazard*. New York: Delta.

Altheide, D. 1992. Gonzo justice. *Symbolic Interaction* 15 (1): 69–86.

Andrée, P. 2002. The biopolitics of genetically modified organisms in Canada. *Journal of Canadian Studies* 37 (3): 162–91.

Anon. 1994. Inside the Royal Commission. In *Misconceptions: The social construction of choice and the new reproductive and genetic technologies*, ed. G.Basen, M. Eichler, and A. Lippman, 223–36. Prescott, ON: Voyageur Publishing.

Aoki, K. 1993a. Authors, inventors and trademark owners: Private intellectual property rights and the public domain. Part 1. *Columbia-VLA Journal of Law & Arts* 18:1–73.

– 1993b. Authors, inventors and trademark owners: Private intellectual property rights and the public domain. Part 2. *Columbia-VLA Journal of Law & Arts* 18:191–267.

Appleby, B.M. 1999. *Responsible parenthood: Decriminalizing contraception in Canada*. Toronto: University of Toronto Press.

Armstrong, J. 1989. Dodd now 'regrets' her abortion: Says pro-choice groups used her. *Toronto Star*, 19 July.

Asbell, B. 1995. *The pill: A biography of the drug that changed the world*. New York: Random House.

Associated Press. 2003. CSI: Real world debunking some of the CSI bunk. *Unknown News*, http://www.unknownnews.net/030909csi.html (accessed 25 May 2005).
– 2005. New case of mad cow in Canada. January 12.
Atwood, M. 1998. *The handmaid's tale*. Toronto: Seal Books.
– 2003. *Oryx and Crake*. Toronto: Seal Books.
– 2004. *The handmaid's tale* and *Oryx and Crake* in context. *PMLA* 119 (3): 513–17.
Avery, D. 2002. Pandora's box of plagues. *Globe and Mail*, 12 August.
Baird, P. 1996. Opinion. *Calgary Herald*, 9 December.
– 1998. Patenting and human genes. *Perspectives in Biology and Medicine* 41 (3): 391–408.
Basen, G., M. Eichler, and A. Lippman, eds. 1994. *Misconceptions: The social construction of choice and the new reproductive and genetic technologies*. Prescott, ON: Voyageur Publishing.
Baudrillard, J. 2002. *The spirit of terrorism*. London: Verso.
Bauman, Z. 2001. War of the globalization era. *European Journal of Social Theory* 4 (1): 11–28.
Beck, U. 1992. *Risk society*. Thousand Oaks: Sage.
– 2002. The terrorist threat: World risk society revisited. *Theory, Culture & Society* 19 (4): 39–55.
Beltrame, J. 1989. Courts busy as Ottawa falls silent on abortion. *Montreal Gazette*, 15 July.
Berkowitz, A., and D.J. Kevles. 2002. Patenting human genes: The advent of ethics in the political economy of patent law. In *Who owns life?* ed. D. Magnus, A. Caplan, and G. McGee, 75–97. Amherst, NY: Prometheus Books.
Bindman, S. 1988. Stage set for easier abortions. *Ottawa Citizen*, 29 January, A1.
Black, D. 2003. Who's afraid of the big, bad ... whatever it is? *Toronto Star*, 2 May, D1.
Bogard, W. 1996. *The simulation of surveillance: Hypercontrol in telematic societies*. Cambridge: Cambridge University Press.
Bolan, K. 1990. Pro-choicers plan caravan to Ottawa. *Vancouver Sun*, 23 April.
Bowring, F. 2003. *Science, seeds and cyborgs: Biotechnology and the appropriation of life*. London: Verso.
Boyd, S. 2000. Campus mice scampering north work for DuPont. *Toronto Star*, 18 August.
Brown, B., N. Miller-Chenier, and S. Norris. 2003. Committees as agents of public policy: The Standing Committee on Health. *Canadian Parliamentary Review* 26 (3): 4–8.
Bunton, R., and A. Peterson, eds. 2005. *Genetic governance: Health, risk and ethics in a biotech era*. New York: Routledge.

Bush, C. 2003. Hex on the city. *Globe and Mail*, 26 April.

Calamai, P. 2002. Findings could aid fight on infection, bio-bugs. *Toronto Star*, 13 July, A11.

Calgary Herald. 2000. Editorial. 5 August.

Cameron, J. 1992. Reproductive choice. *Globe and Mail*, 22 February.

Campbell, J. 1996. Genetically engineered mice at centre of debate. *Ottawa Citizen*, 24 February.

Canada NewsWire. 2001. Supreme Court hearing of Harvard mouse case detrimental to life-saving innovations. June 20.

Canadian Biotechnology Advisory Committee. 2001a. *Biotechnological intellectual property and the patenting of higher life forms: Consultation document 2001*. Ottawa: Canadian Biotechnology Advisory Committee.

– 2001b. *Interim report on biotechnology and intellectual property*. Ottawa: Canadian Biotechnology Advisory Committee.

– 2002a. *Patenting of higher life forms*. Ottawa: Canadian Biotechnology Advisory Committee.

– 2002b. *Summary of Responses*. Ottawa: Canadian Biotechnology Advisory Committee.

Canadian Police Association. 1998. *Brief to the Standing Committee on Legal and Constitutional Affairs regarding Bill C-3*. Ottawa: Canadian Police Association.

Canadian Press. 2002. Ontario MPP unveils police bioterror gear. *Globe and Mail*, 12 July.

Canadian Public Health Association. 2001. *Animal-to-human transplantation: Should Canada proceed? A public consultation on xenotransplantation*. Ottawa: Canada.

Carey, E. 2005. Ebola vaccine breakthrough; Vaccines may be in use in 5 years; Winnipeg facility hailed for strides on two killer viruses; Possibility of bio-terrorism attack was a strong factor, MD says. *Toronto Star*, 6 June.

Castel, R. 1991. From dangerousness to risk. In *The Foucault effect: Studies in governmentality*, ed. G. Burchell, C. Gordon, and P. Miller, 281–98. Chicago: University of Chicago Press.

CBC Television Prime Time News. 1994. Broadcast transcripts. 18 August (accessed from Canadian NewsStand 6 March 2004).

CBSNEWS.com. 2003. 'CSI' spurs forensic academics. 18 August. http://www.cbsnews.com/stories/2003/08/18/entertainment/printable 568982.shtml.

Chang, K. 2003. Researchers seek better sensors; Bioterror fears spur efforts to improve detectors; Fast, accurate, simple, cheap systems needed. *Toronto Star*, 13 April.

Chase, S. 2006. U.S. plans tougher inspections at border. *Globe and Mail*, 1 September.

Chatelaine. 1990. Chatelaine's newsmaker of the year '90. January, 38–42, 113.

Chidley, J. 2002. Of mice and men. *Canadian Business,* 30 December, 8.

Christie, J. 2000. Should Canada allow patents on new plants and animals? *London Free Press,* 3 January.

Clute, C. 2002. Reveal sperm's secrets. *Ottawa Citizen,* 14 June.

Committee on the Operation of the Abortion Law. 1977. *Report of the Committee on the Operation of the Abortion Law.* Catalogue no. J2-30/1977. Ottawa: Supply and Services Canada.

Cook, R. 1999. *Vector.* New York: Putnam's.

Coombe, R. 1992. The celebrity image and cultural identity: Public rights and the subaltern politics of gender. *Discourse* 14 (3): 59–88.

Dagognet, F. 1988. *La Maîtrise du vivant.* Paris: Hachette.

Davies, M., and N. Naffine. 2001. *Are persons property? Legal debates about property and personality.* Aldershot: Ashgate.

de Lint, W., and S. Virta. 2004. Security in ambiguity: Towards a radical security politics. *Theoretical Criminology* 8 (4): 465–89.

Deonandan, R. 2003. Is SARS a precursor to pandemic? *Toronto Star,* 26 May.

Drexler, M. 2002. *Secret agents: The menace of emerging infections.* Washington, DC: Henry.

Duffield, M. 2001. *Global governance and the new wars: The merging of development and security.* London: Zed Books.

Dumit, J. 2004. *Picturing personhood: Brain scans and biomedical identity.* Princeton, NJ: Princeton University Press.

Edmonton Journal. 1989. Dodd switches to pro-life after battle to have abortion. 19 July.

– 1992. Women's group blasts commission. *Edmonton Journal,* February 4, C8.

Eichler, M. 1994. Frankenstein meets Kafka: The Royal Commission on New Reproductive Technologies. In *Misconceptions: The social construction of choice and the new reproductive and genetic technologies,* ed. G. Basen, M. Eichler, and A. Lippman, 196–222. Prescott, ON: Voyageur Publishing.

Einsiedel, E., and F. Timmermans, eds. 2005. *Crossing over: Genomics in the public arena.* Calgary: University of Calgary Press.

Einsiedel, E.F., and L. Goldberg. 2004. Dwarfing the social? Nanotechnology lessons from the biotechnology front. *Bulletin of Science, Technology & Society* 24 (1): 28–33.

Eisenberg, R.S. 2002. How can you patent genes? In *Who owns life?* ed. D. Magnus, A. Caplan, and G. McGee, 117–34. Amherst, NY: Prometheus Books.

Environics Research Group. 2000. Canadian biotechnology community's practices, attitudes and opinions regarding the research exemption and methods of medical treatment exemption in Canadian patent law. Draft report.

Etkowitz, H., and A. Webster. 1995. Science as intellectual property. In *Handbook of science and technology studies*, ed. S. Jasanoff, E. Markle, J.C. Petersen, and T. Pinch, 480–505. London: Sage.

Fegan, E.V. 1996. Fathers' foetuses and abortion decision-making: The reproduction of maternal ideology. *Social & Legal Studies* 5:75–93.

Fidelman, C. 1996. Bill focuses debate on ethics of reproduction. *Montreal Gazette*, 4 November.

Ford, T. 1998. There may be some good things about cloning humans. *Toronto Star*, 30 July.

Foucault, M. 1986. *An introduction*. Vol. 1 of *The history of sexuality*. Trans. Robert Hurley. New York: Pantheon Books.

– 1988. *The care of the self*. Vol. 3 of *The history of sexuality*. New York: Vintage.

– 2003. *Society must be defended: Lectures at the Collège de France 1975–6*. Ed. M. Bertani and A. Fontana. New York: Picador.

Fox, M. 2000. Pre-persons, commodities or cyborgs: The legal construction and representation of the embryo. *Health Care Analysis* 8:171–88.

Fox Keller, E. 2000. *The century of the gene*. Cambridge, MA: Harvard University Press.

Freeze, C. 2001. Would you allow patents for life forms? *Globe and Mail*, 23 March.

Frketich, J. 2004. Hamilton MD to lead attack on West Nile virus. *Toronto Star*, 12 November.

Fukuyama, F. 1989. The end of history? *National Interest* (Summer): 3–18.

Geddes J., S. Mcclelland, and P. Chisholm. 1999. Making babies: In the age of in vitro fertilization does the state have a place in the test tubes of the nation? *Maclean's*, 6 December.

Gerlach, N. 2004. *The genetic imaginary: DNA in the Canadian criminal justice system*. Toronto: University of Toronto Press.

Gerlach, N., and S.N. Hamilton. 2003. Introduction: A history of social science fiction. *Science Fiction Studies* 30 (2): 161–73.

– 2005. From mad scientist to bad scientist: Richard Seed as biogovernmental event. *Communication Theory* 15 (1): 78–99.

Gerstel, J. 2003a. The good, the bad and the uh-oh. *Toronto Star*, 4 April, D3.

– 2003b. WHO knows what's happening. *Toronto Star*, 25 April, F1.

– 2004. Flu pandemic feared; Experts 'very worried' about Asian bird flu mutating into a contagious, deadly human virus. *Toronto Star*, 5 November, C1.

Gervais, D. 2002. Eek! The mouse confounds the House! *Ottawa Citizen*, 14 December.

Giddens, A. 1985. *The nation-state and violence*. Vol. 2 of *A contemporary critique of historical materialism:* Berkeley: University of California Press.

Globe and Mail. 1988. Rejecting the rules. 29 January.

– 1993. Proceed with care indeed. 2 December.

– 2007. Editorial. 29 December.

Gold, R. 2000. *Patents in genes.* Ottawa: Canadian Biotechnology Advisory Committee.

Graham, E.L. 2002. *Representations of the post/human: Monsters, aliens and others in popular culture.* Manchester: Manchester University Press.

Hall, N. 1994. Rapist linked to Milgaard case walks free today. *Vancouver Sun,* 26 May.

Hamel, R.M. 2002. Cloning inaccurately defined in Bill C-56. *Cornwall Standard-Freeholder,* 11 June.

Hamilton, S.N. 2003. Traces of the future: Biotechnology, science fiction, and the media. *Science Fiction Studies* 30 (2): 267–73.

Hamilton Spectator. 2002. Embryos are human life. 13 June.

Hanson, M.J. 2002. Patenting genes and life: Improper commodification. In *Who owns life?* ed. D. Magnus, A. Caplan, and G. McGee, 161–74. Amherst, NY: Prometheus Books.

Haraway, D.J. 1997. *Modest_Witness@Second_Millennium. FemaleMan©_Meets_OncoMouse™: Feminism and technoscience.* New York: Routledge.

Hardt, M., and A. Negri. 2000. *Empire.* Cambridge, MA: Harvard University Press.

– 2005. *Multitude: War and democracy in the age of empire.* New York: Penguin.

Harper, T. 2002. Ottawa orders 10 million anti-smallpox doses. *Toronto Star,* 28 November.

Hartouni, V. 1997. *Cultural conceptions: On reproductive technologies and the remaking of life.* Minneapolis: University of Minnesota Press.

Health Canada. 2002. *Congenital anomalies in Canada: A perinatal health report, 2002.* Ottawa: Health Canada.

Healy, P. 1995. Statutory prohibitions and the regulation of new reproductive technologies under federal law in Canada. *McGill Law Journal* 40:905–46.

Hempel, C. 2003. TV's whodunit effect. *Boston Globe Magazine,* 2 September.

Hrabluk, L. 2002. Harvard mouse case shows need to update Patent Act of 1869. *New Brunswick Telegraph Journal,* 22 May.

Humphreys, A. 2003. Unless it's TV, murder can be dull work. *National Post,* 25 February.

Huntington, S. 1997. *The clash of civilizations and the remaking of world order.* New York: Touchstone.

Hurst, L. 1991. If it becomes a battleground … *Toronto Star,* 1 December.

Hutchinson, E. 2000. Letter to the editor. *Ottawa Citizen,* 24 August.

Immen, W. 2002. Another deadly plague possible, study says. *Globe and Mail,* 26 April.

International Clearinghouse for Birth Defects Monitoring Systems. 2003. *Annual Reports 2003*. Rome: International Centre on Birth Defects.

Ishiguro, K. 2005. *Never let me go*. New York: Knopf.

Jang, B. 1994. Report clears Saskatchewan's handling of Milgaard; Probe found no evidence to back convicted killer's claims to innocence. *Montreal Gazette*, 17 August.

Jeffs, A. 1993. Reproductive technology report may fuel new storm: Recommendations to provide guide for rules. *Ottawa Citizen*, 9 November.

Jenish, D. 1989. Abortion on trial: The debate takes on a personal edge as it moves into the courts. *Maclean's*, 31 July, 14–17.

Jha, P. 2003. Not so fast, Toronto. *Globe and Mail*, 30 April.

Joseph, A., and A. Winter. 1996. Making the match: Human traces, forensic experts and the public imagination. In *Cultural babbage: Technology, time and invention*, ed. F. Spufford and J. Uglow, 193–214. London: Faber and Faber.

Kalla, D. 2005. *Pandemic*. New York: TOR-Forge.

Karp, C., and C. Rosner. 1991. The Milgaard story: A body in the snow, a life in penitentiary. *Vancouver Sun*, 12 December.

Kay, L.E. 2000. *Who wrote the book of life? A history of the genetic code*. Stanford: Stanford University Press.

Keim, P. 2002. Bioterrorism fears help propel research efforts. *Toronto Star*, 2 January.

Kellner, D., and S. Best. 2001. *The postmodern adventure: Science, technology, and cultural studies at the third millennium*. New York: Guilford.

Kenney, M. 2007. *From Pablo to Osama: Trafficking and terrorist networks, government bureaucracies, and competitive adaptation*. University Park, PA: Pennsylvania State University Press.

Klein, G. 2002. Top court deals blow to biotech industry. *Star-Phoenix*, 6 December.

Kirkey, S. 2002. A better mouse rejected. *National Post*, 6 December.

Koring, P. 2002. On the trail of the anthrax killer. *Globe and Mail*, 6 March.

Laing, B. 1989. Guidelines for abortion needed. *Globe and Mail*, 15 July.

Laver, R. 1989. The debate about life. *Maclean's*, 31 July, 20.

Leblanc, D. 2003. Time to beef up food safety, experts say. *Globe and Mail*, 22 May.

Lipson, K. 2004. The 'Nerd Squad': Forensic examiners are TV's new high-tech heroes. *Newsday.com*. http://www.newsday.com/entertainment/ny-ent-foren 924,0,7850561.story (accessed 25 May 2005).

Low, D. 2003. And now for the bad news. *Globe and Mail*, 9 August.

Lu, V. 2003. City's trash trucks face inspections. *Toronto Star*, 5 April.

Lyon, D. 2001. *Surveillance society: Monitoring everyday life*. Philadelphia: Open University.

MacQueen, K. 1988. Ruling on abortion reflects major shift in power sharing: Rights Charter boosts role of judiciary. *Montreal Gazette*, 30 January.

Manning, P. 2002. We're hatching a flawed bill. *Globe and Mail*, 27 May.

Marchessault, J. 2000. David Suzuki's *The secret of life*: Informatics and the popular discourse of the life code. In *Wild science: Reading feminism, medicine and the media*, ed. J. Marchessault and K. Sawchuk, 55–65. New York: Routledge.

Marchessault, J., and K. Sawchuk, eds. 2000. *Wild science: Reading feminism, medicine and the media*. New York: Routledge.

Marr, J.S. and J. Baldwin. 1998. *The eleventh plague.* New York: Cliff Street Books.

Mason, D. 2003. CSI and the real world. *Halifax Daily News*, 17 February.

McClearn, M. 2000. Of mice and men: With the second-largest biotech industry in the world, why aren't we talking about gene patenting? *Canadian Business* 73 (19): 129.

McIlroy, A. 2002. Killer enzyme could combat bioterror. *Globe and Mail*, 22 August.

McKeague, P. 1991. Media ignoring the big picture. *Windsor Star*, 18 December.

McLaren, A. 1999. *Twentieth-century sexuality: A history.* Oxford: Blackwell.

McLaren, A., and A.T. McLaren. 1986. *The bedroom and the state: The changing practices and politics of contraception and abortion in Canada, 1880–1990.* Toronto: McLelland and Stewart.

McTeer, M. 1996. Put harness on power over life. *Calgary Herald*, 28 June, A18.

– 1999. *Tough choices: Living and dying in the twenty-first century.* Toronto: Irwin Law.

– 2000. Moral lessons from a mouse. *Ottawa Citizen*, 9 August, A15.

Merz, J.F. 2002. Discoveries: Are there limits on what may be patented? In *Who owns life?* ed. D. Magnus, A. Caplan, and G. McGee, 99–116. Amherst, NY: Prometheus Books.

Mitchell, R. 2004. $ell: Body wastes, information, and commodification. In *Data made flesh: Embodying information*, ed. R. Mitchell and P. Thurtle, 121–36. New York: Routledge.

Morton, F.L. 1992. *Morgentaler v. Borowski: Abortion, the Charter and the courts.* Toronto: McClelland and Stewart.

Moysa, M. 1996. Doctors critical of government plan: Reproductive bill panned. *Edmonton Journal*, 22 October.

Münkler, H. 2005. *The new wars.* Cambridge: Polity.

Naffine, N. 1998. The legal structure of self-ownership: Or the self-possessed man and the woman possessed. *Journal of Law and Society* 25 (2): 193–212.

National Advisory Committee on SARS and Public Health. 2003. *Learning from SARS: Renewal of public health in Canada.* Ottawa: Health Canada.

Nelkin, D., and M.S. Lindee. 1995. *The DNA mystique: The gene as cultural icon.*
New York: Freeman.

Neocleous, M. 2000. Against security. *Radical Philosophy* 100:7–15.

New Hampshire Police Standards and Training Council. 2000. *Articulable
suspicion* (electronic edition). October.

Nichols, M., and S.D. Driedger. 1993. Banned parenthood: Report by the Royal
Commission on New Reproductive Technologies. *Maclean's*, 13 December,
52.

Novas, C., and N. Rose. 2000. Genetic risk and the birth of the somatic individual. *Economy and Society* 29 (4): 485–513.

O'Mahony, P. 1999. *Nature, risk and responsibility: Discourses of biotechnology.* New
York: Routledge.

Order in Council P.C. 2150, 25 October 1989.

Ottawa Citizen. 1993. Reproductive technology. 30 November.

– 1997. Longer gestation required. 21 March.

– 2000. Editorial. 11 August.

– 2002. The mouse trapped: Canada must fix its patent law or we'll be left
behind in biotech. 6 December.

Page, S. 1993. Take profits out of births. *Kitchener-Waterloo Record,* 30 November.

Pal, L.A. 1991. How Ottawa dithers: The Conservatives and abortion policy. In
How Ottawa spends: The politics of fragmentation 1991–92, ed. F. Abele, 269–306.
Ottawa: Carleton University Press.

Palmer, K. 2003. Trio maps behaviour of elusive Ebola virus. *Toronto Star,*
11 April.

Paterson, A.K. 1989. Without law, judgments meaningless. *Montreal Gazette,*
21 July.

Plischke, H. 1995. The blooding of a prairie town. *Ottawa Citizen,* 13 May.

– 1996a. Process of elimination: Mass DNA testing in hunt for rapist. *Edmonton
Journal,* 14 June.

– 1996b. Vermilion rapist nocturnal, loner, RCMP tells 200. *Edmonton Journal,*
19 June.

Poster, M. 1996. Databases as discourse: Or, electronic interpellations. In
Computers, surveillance, and privacy, ed. D. Lyon and E. Zureik, 175–92.
Minneapolis: University of Minnesota Press.

Pottage, A. 1998. The inscription of life in law: Genes, patents, and bio-politics.
In *Law and human genetics: Regulating a revolution,* ed. R. Brownsword, W.R.
Cornish, and M. Llewelyn, 148–73. Oxford: Hart.

Preston, D., and L. Child. 1996. *Mount Dragon.* New York: Tor.

Preston, R. 1997. *The cobra event.* New York: Random House.

– 2002. *The demon in the freezer.* New York: Random House.

Privacy Commissioner of Canada. 1998. *Remarks to the Standing Committee on Justice and Human Rights Concerning Bill C-3, the DNA Identification Act.* Ottawa: Privacy Commission of Canada.

Privy Council Office. 2004. *Securing an open society: Canada's national security policy.* Ottawa: Government of Canada.

Rabinow, P. 1996. *Essays on the anthropology of reason.* Princeton, NJ: Princeton University Press.

Rabinow, P., and N. Rose. 2003. Thoughts on the concept of biopower today. Unpublished paper.

Recer, P. 2002. Terrorist strain of anthrax traced to Texan ranch. *Globe and Mail,* 2 February.

ReGenesis. 2004–5. Television series, episodes 1–13. United States: The Movie Network.

Regis, E. 1999. *The biology of doom: The history of America's secret germ warfare project.* New York: Owl.

Robertson, G., M. Marsh, L. Tickner, J. Bird, B. Curtis, and T. Putnam, eds. 1996. *Futurenatural: Nature/science/culture.* New York: Routledge.

Roof, J. 2007. *The Poetics of DNA.* Minneapolis: University of Minnesota Press.

Rose, N. 2001. The politics of life itself. *Theory, Culture & Society* 18 (6): 1–30.

Ross, R. 2002. Biotech called to duty in the U.S. war on terrorism: Countering germ warfare is major part of strategy. *Toronto Star,* 13 June.

Royal Commission on New Reproductive Technologies. 1993. *Proceed with care: Final report of the Royal Commission on New Reproductive Technologies; Summary and highlights.* Ottawa : Canada Communications Group.

– 1993. *Social values and attitudes surrounding new reproductive technologies.* Ottawa: Canada Communications Group.

Ruhl, P.L. 2002. Disarticulating liberal subjectivities: Abortion and fetal protection. *Feminist Studies* 28 (1): 37–60.

Sallot, J. 2002. Iraq's bioterrorism labs pose biggest threat, expert warns. *Globe and Mail,* 17 September.

The SARS Commission (Ontario). 2006. *The SARS Commission executive summary.* Toronto: Government of Ontario.

Sawchuk, K. 2000. Biotourism, *Fantastic Voyage,* and sublime inner space. In *Wild science: Reading feminism, medicine and the media,* ed. J. Marchessault and K. Sawchuk, 9–23. New York: Routledge.

Schabas, R. 2005. Much ado about clucking. *Globe and Mail,* 8 October.

Seide, R.K., and C.L. Stephens. 2002. Ethical issues and application of patent laws in biotechnology. In *Who owns life?* ed. D. Magnus, A. Caplan, and G. McGee, 59–73. Amherst, NY: Prometheus Books.

Shaw, M. 2005. *The new western way of war.* Cambridge: Polity.

Shelton, D.E., Y.S. Kim, and G. Barak. 2006. A study of juror expectations and demands concerning scientific evidence: Does the 'CSI effect' exist? *Vanderbilt Journal of Entertainment and Technology Law* 9 (2): 330.

Shepard, C. 2001. CBS' hot new drama CSI ... The good, the bad, and the ugly. *Young Forensic Scientists Forum Newsletter*, January.

Shephard, M. 2003. Are we prepared for bioterror attack? Reaction to SARS outbreak being analyzed: Cracks seen, Canada's readiness in doubt. *Toronto Star*, 5 April.

– 2004. U.S. better prepared for smallpox, expert says; 9/11 raised timely warning about bioterrorism; Vaccine stocks boosted to 300 million doses. *Toronto Star*, 21 October.

Shiva, V. 1997. *Biopiracy: The plunder of nature and knowledge.* Boston: South End.

– 2005. *Earth democracy: Justice, sustainability and peace.* Cambridge, MA: South End.

Simpson, J. 1988. No easy choice. *Globe and Mail*, 29 January.

Sixth Review Conference of the States Parties to the Convention on the Prohibition on the Development, Production and Stockpiling of Bacteriological (Biological) and Toxing Weapons and on Their Destruction. 2006. *Final document.* Geneva.

Smith, D. 1988. Ruling 'uncivilized,' cardinal says. *Toronto Star*, 29 January.

Squier, S. 1998. Interspecies reproduction: Xenogenic desire and the feminist implications of hybrids. *Cultural Studies* 12:360–81.

Star Wire Services. 2002. Feat raises bioterror fears. *Toronto Star*, 12 July.

Starr, R. 2003. Biotechnology. *Canadian Business* 76 (1): 38.

Statistics Canada. 1999. General social survey. Cycle 13..Victimization. Ottawa: Statistics Canada.

Strauss, S. 1996. News from nowhere, policy to follow: Media and the social construction of three strikes and you're out. In *Three strikes and you're out: Vengeance as public policy*, ed. D. Shichor and D. Sechrest, 177–202. Thousand Oaks: Sage.

– 1998. Legal spin on DNA sways jury, studies say. *Globe and Mail*, 17 February.

Strydom, P. 1996. The civilization of the gene: Biotechnological risk framed in the responsibility discourse. In *Nature, risk and responsibility: Discourses on biotechnology*, ed. P. O'Mahony, 21–36. New York: Routledge.

Sullivan, P. 2003. It's a small world when disease strikes. *Globe and Mail*, 12 May.

Sullivan, R. 2005. An embryonic nation: Life against health in Canadian biotechnological discourse. *Communication Theory* 15 (1): 39–58.

Talaga, T. 2005. Canada looks at anthrax vaccine; Firm supplying U.S. in discussions with Ottawa; American initiative to protect against bioterror. *Toronto Star*, 21 April.

Thacker, E. 2005. *The global genome: Biotechnology, politics, and culture.* Cambridge, MA: MIT Press.

Tibbetts, J. 2000. Animals can be patented, Court of Appeal rules. *Calgary Herald,* 4 August.

Tobin, A.M. 2006. Bioterror testing: Apes survive virus with vaccine given later. *Globe and Mail,* 27 April.

Tone, A. 2001. *Devices & desires: A history of contraceptives in America,* New York: Hill and Wang.

Trigueiro, D. 1996. Law tries to limit social costs of reproductive technology. *Calgary Herald,* 7 July.

Turney, J. 1998. *Frankenstein's footsteps: Science, genetics and popular culture.* New Haven: Yale University Press.

Tyler, T. 1997. DNA clears Milgaard. *Toronto Star,* 19 July.

Tyler, T.R. 2006. Viewing *CSI* and the threshold of guilt: Managing truth and justice in reality and fiction. *Yale Law Journal* 115:1050–85.

United Nations. 2004. *A more secure world, our shared responsibility: Report of the High Level Panel on Threats, Challenges and Change.* New York: United Nations.

United Nations General Assembly. 2006. *Uniting against terrorism: Recommendations for a global counter-terrorism strategy.* New York: United Nations.

Vancouver Province. 2002. CSI TV shows force coroners to dig deeper. 1 December.

Vancouver Sun. 1992. Panel too close to pharmaceutical firms, group says. 4 February.

– 1993. Of eggs, sperm, morality and technology. 29 December.

– 2002. Editorial. 7 December.

Vandelac, L. 1994. The Baird Commission: From 'access' to 'reproductive technologies' to the 'excesses' of practitioners or the art of diversion and relentless pursuit. In *Misconceptions: The social construction of choice and the new reproductive and genetic technologies,* ed. G. Basen, M. Eichler, and A. Lippman, 253–72. Prescott, ON: Voyageur.

van der Ploeg, I. 2002. Biometrics and the body as information: Normative issues in the socio-technical coding of the body. In *Surveillance as social sorting: Privacy, risk, and automated discrimination,* ed. D. Lyon, 57–73. New York: Routledge.

van Dijck, J. 1995. *Manufacturing babies and public consent: Debating new reproductive technologies.* London: MacMillan.

– 1998. *ImagEnation: Popular images of genetics.* New York: New York University Press.

van Dusen, L. 1989. Sex, law and ethics: An inquiry into the technologies of life. *Maclean's*, 13 November, 16.

Vienneau, D. 1988a. 'Business as usual' on abortions doctors told. *Toronto Star*, 29 January.

– 1988b. Women must find physician willing to perform abortions. *Toronto Star*, 29 January.

Wahl, A. 2002. The case of the Harvard mouse: Should scientists be allowed to patent higher life-forms? That's up to the Supreme Court of Canada. *Canadian Business* 75 (16): 75.

Waldby, C. 2000. *The visible human project: Informatic bodies and posthuman medicine*. London: Routledge.

Walkom, T. 1997. Federal Court to rule on patenting higher life forms. *Toronto Star*, 18 November.

Wallace, B., and L. van Dusen. 1989. Abortion agony. *Maclean's*, 21 August, 16.

Walmsley, A. 1989. A change of heart: Dodd's regrets cause family turmoil. *Maclean's*, 31 July, 18–19.

Walzer, M. 1977. *Just and unjust wars: A moral argument with historical illustrations*. New York: Basic Books.

Warrick, J., and J. Mintz. 2003. Deadly secrets in a toothpaste tube. *Toronto Star*, 21 April.

Watkins, E.S. 1998. *On the pill: A social history of oral contraceptives 1950–70*. Baltimore: Johns Hopkins University Press.

Watters, W.W. 1976. *Compulsory parenthood: The truth about abortion*. Toronto: McClelland and Stewart.

Wayt Gibbs, W. 2002. Biological advances carry their own risks. *Toronto Star*, 1 March.

Weston, G. 1989. Father's rights: Are they entitled to a say about the fetus? *Ottawa Citizen*, 21 July.

Wigod, R. 1993. Genetic genie is out of the bottle: Maureen McTeer says federal laws needed immediately to help women, kids. *Vancouver Sun*, 8 February.

Willing, R. 2004. CSI effect has juries wanting more evidence. *USA Today*, 5 August.

Wood, C. 1996. Beyond abortion: Advances in science leave an old debate in the dust. *Maclean's*, 19 August, 14–15.

World Health Organization. 2002. *Preparedness for the deliberate use of biological agents: A rational approach to the unthinkable*. Geneva: World Health Organization.

Yelaja, P. 2002. Scientists renew call for biohazard lab: Etobicoke site should be reopened in wake of Sept. 11. *Toronto Star*, 25 January.

Zuboff, S. 1988. *In the age of the smart machine: The future of work and power.* New York: Basic Books.

Films

The Andromeda Strain. Dir. and Prod. R. Wise. 1971. Universal Pictures.
Gattaca. Dir. and screenwriter A. Niccol. Prod. D. Devito. 1997. Columbia Pictures.
The Island. Dir. M. Bay. Prod. M. Bay and K. Bates. 2005. Dreamworks SKG.
Outbreak. Dir. W. Petersen. Prod. D. Henderson and A. Kopelson. 1995. Warner Brothers.
The Sixth Day. Dir. R. Spottiswoode. Prod. M. Medavoy, A. Schwarzenegger, and J. Davison. 2000. Columbia Tristar.
28 Days Later. Dir. D. Boyle. Prod. A. MacDonald. 2002. 20th Century Fox.
28 Weeks Later. Dir. J.C. Fresnadillo. Prod. D. Boyle and A. Garland. 2007. Fox Atomic.

Legislation

An Act to Amend the Criminal Code and the Young Offenders Act (forensic DNA analysis) R.S.C 1995, c. 27 (Bill C-104).
Anti-terrorism Act S.C. 2001, c. 41.
Assisted Human Reproduction Act S.C. 2004, c. 2.
Canada Health Act R.S.C. 1985, c. C-6.
Canadian Charter of Rights and Freedoms S.C. 1982, c. 11.
Criminal Code R.S.C. 1985, c. C-46.
DNA Identification Act S.C. 1998, c. 37.
Emergency Management Act S.C. 2007, c. 15.
Health Protection and Promotion Act R.S.O. 1990, c. H.7.
Human Reproduction and Genetic Technologies Act, Bill C-47, 2nd Session, 35th Parliament 45–46 Elizabeth II 1996–97.
The Patent Act, R.S.C. 1985, c. P-4.
Public Safety Act, 2002, S.C. 2004, c. 15.

Cases

Borowski v. Attorney General of Canada and Minister of Finance of Canada, 1983 (Sask. Q.B.).
Borowski v. Attorney General of Canada and Minister of Finance of Canada, 1987 (Sask. C.A.).

Borowski v. Attorney General of Canada and Minister of Finance of Canada, 1989 (S.C.C.).

Commissioner of Patents (M. Leesti). (1995). *Decision of the Commissioner of Patents, No. 1203 (4 August).* Hull, Quebec: Canadian Intellectual Property Office.

Diamond v. Chakrabarty (1980) 447 US 303 (S.C. of U.S.).

Harvard College v. Canada (Commissioner of Patents) (1998) 79 C.P.R. (3d) 98 (F.C.T.D.).

Harvard College v. Canada (Commissioner of Patents) (2000) 7 C.P.R. (4th) 1 (F.C.A).

Harvard College v. Canada (Commissioner of Patents) (2002) 21 C.P.R. (4th) 417 (S.C.C.).

Monsanto Canada Inc. v. Schmeiser (2001) 12 C.P.R. (4th) 204 (F.C.T.D.).

Monsanto Canada Inc. v. Schmeiser (2002) 21 C.P.R. (4th) 1 (F.C.A).

Monsanto Canada Inc. v. Schmeiser (2004) 31 C.P.R. (4th) 161 (S.C.C.).

Moore v. The Regents of The University of California et al. (1990) 51 Cal. (3d) 120 (S.C. of Cal.).

Murphy v. Dodd, 1989a (Ont. H.C.J.).

Murphy v. Dodd, 1989b (Ont. S.C.).

Pioneer Hi-Bred v. Commisioner of Patents (1989) 25 C.P.R. (3d) 257 (S.C.C.).

R. v. Borden (1994) 24 C.R.R. (2d) 51 (S.C.C.).

R. v. Bourguignon (1991) OJ 2670.

R. v. Dyment (1988) 45 C.C.C. (3d) 244 (S.C.C.).

R. v. Morgentaler (1975) 20 C.C.C. (2d) 449 (S.C.C.).

R. v. Morgentaler (1988) 62 C.R. (3d) 1 (S.C.C.).

Re Application for Patent of Abitibi Co. (1982) 62 C.P.R. (2d) 81 (Patent Appeal Board).

Rodriguez v. British Columbia (Attorney General) (1993) 24 C.R. (4th) 281 (S.C.C.).

Tremblay v. Daigle, 1989a (Quebec Sup. Ct).

Tremblay v. Daigle, 1989b (Que. C.A.).

Tremblay v. Daigle, 1989c, [1989] 2 S.C.R. 530, 62 D.L.R. (4th) 634, 102 N.R. 81 (S.C.C.).

Index